ECONOMIC POLITICAL GEOGRAPHY

ECONOMIC POLITICAL GEOGRAPHY

By

Dr. Pradeep Sharma

DISCOVERY PUBLISHING HOUSE
NEW DELHI-110002

First Published - 2007

Reprinted - 2017

ISBN: 978-81-8356-196-9

Economic Political Geography

Published by:

DISCOVERY PUBLISHING HOUSE PVT. LTD.
4383/4B, Ansari Road, Darya Ganj
New Delhi-110 002 (India)
Phone: +91-11-23279245, 43596064-65
Fax: +91-11-23253475
E-mail: discoverypublishinghouse@gmail.com
sales@discoverypublishinggroup.com
web: www.discoverypublishinggroup.com

Printed at:
Infinity Imaging Systems
Delhi

PREFACE

This book Economic Political Geography has been written for the use of the students of Degree, Honours and Postgraduate classes. The subject matter of this book has been discussed in such a simple way that the student find no difficulty to understand it. The historical perspective does not, however, conflict with the premise that the old order must change yielding place to new, it only help us to identify better the newness of the new. As times and technology change the entire perspective of society on its environment experiences change which in turn, influences our value systems and form of social and political organisation.

The author is thankful to all those who have assisted him in one way or the other in the preparation of the book. He would like to express his gratefulness to the scholars whose book and article have provided the necessary source material for this book.

It is hoped that the book will be useful to student of economic political geography and also to those appearing at various competitive examinations.

Author

CONTENTS

1

ECONOMIC ELEMENTS OF POLITICAL GEOGRAPHY AND STATE TERRITORIAL PHENOMENON

The states and the relations between them are greatly influenced by the economic elements of political geography. The economic elements are comprised of the basic natural resources of food, power, minerals etc., industry, trade and transportation. These six elements and their effect on the economic and political development of the state will be described and analysed in the following pages.

FOOD RESOURCES

As food supply is necessary for human welfare, it is the primary duty of each state to provide adequate food to its population. Human food consists of cereal and pulses, fruits and vegetables, milk and its products, meat, fish, eggs, sugar, oils etc. No state can be entirely self-sufficient in all food resources, since it is not possible for it to produce all types of food within its own territory. Nevertheless, the state has great responsibility for the supply of at least the staple food to the people much as cereals and pulses in the case of India. The self-sufficiency in food attained by a state is of course a great achievement but only few states such as the U.S.A. and France perhaps come closest to self-sufficiency in food. The states which are unable to produce the required amount of food must import it from the surplus countries. The industrial nations of Western Europe have long been importing food-stuffs from other countries in exchange for their manufactured goods. But poor and developing nations, whose purchasing power is low, are dependent on food supplies as loan or gift from other nations, particularly at the time of famine and drought. Such help is very often interpreted as interference with the political freedom of the nation involved. It has been said that hungry people may willingly forgo their freedom and accept foreign domination, whether communist or non-communist. "Food, probably the most important of the economic elements, has become a powerful political weapon in a world torn by conflicting philosophies and doctrines".

Too much dependence on food imports is undoubtedly a negative factor in the power of the state. The uninterrupted food imports are possible during peace time. But during the war food supplies may not be forthcoming since the transportation routes are blocked. Sufficient stock of food is therefore necessary in every state for use during the time of emergency. In should be the endeavour of each state to raise the production of food stuffs, so that dependence on imports is reduced and self-sufficiency so necessary for national strength is ultimately achieved.

India's Food Situation

At the time of India's independence, the food situation was extremely critical. In the beginning the country had very heavy dependence on food imports, as home production fell short of the requirement. But as difficulties in importing food specially cereals (wheat and rice) increased, it became necessary to increase foodgrains production in the country. The Government of India has tried to solve the food problem in a planned manner. In each of the Five Year Plans greater emphasis was laid on agricultural development as seventy percent of the people depend on this primary sector of the economy. The Green Revolution helped the Indian farmer in increasing the yield per acre of major food crops. More and more irrigation water was made available to the farmer, who was induced to use high yielding varieties of seeds, fertilizers and modern farming equipment. The result has been quite encouraging. The food situation is now much more improved than what it was at the time of independence. From a small amount of 55 million tons in 1949-50 the foodgrain production increased to 150 million tons in 1985-86. Whereas the population increased by less than two times from 36 crores in 1951 to 68 crores in 1981, the foodgrain production rose by about three times.

India is now almost self-sufficient in foodgrain production. The estimates put the production of foodgrains at 170 million tons in 1988-89. Although, India imported cereals and cereal preparations during 1983-86 she is now in a position to export some wheat and rice to other countries.

Power Resources

Power in one form or the other enters practically every single economic activity of the state and its availability and cost determine the economic future of the nation. Besides nuclear energy other sources of power are coal, oil, natural gas and hydroelectricity. As reserves of coal and oil are being depleted fast, governments have thought of using non-conventional forms of energy such as biogas, solar and wind energy. The extensive use of energy leads to economic prosperity which in turn enhances political power of the state. The home production of power and easy access to foreign soon are important factors in the evaluation of the power of the state.

No state is self-sufficient in power production. What to say of West Europen countries, Japan and India, even the U.S.A. is dependent on oil of the Middle East. Although, energy is in great demand in farm, factory and transport during peace time, its importance increases enormously during the war since easy availability of power, specially oil, is a deciding factor in the outcome of war. The importance of oil is great in modern economic development and a modern war cannot be fought without the use of large amount of petroleum and its products in various forms. "The failure of the German armies in their surge through Russia to occupy the Caucasus oil fields (only the Maikop field was occupied) and the later destruction of the German synthetic oil plant, were two great victories for the Allies. Their significance is evident by the fact that at the end of the war thousands of German planes ware captured on the air fields, unable to get away because they had no oil" (Van Valkenburg, 1963).

India's Power Resources

The coal position of India is satisfactory but the situation with respect to oil is rather alarming. The country has registered a five-time increase in coal production Since independence, from a mere 38 million tons in 1947, the coal production has increased to 154 million tons in 1985-86. According to the Geological Survey of India estimates of 1985, the Indian coal reserves were 1,55,900 million tons for coal seams of 0.5 metre and above in thickness down to depth of 1200 metres. The coal production is almost completely in the public sector, as coal mining was nationalized in 1972 and the Coal India Ltd. came into being in 1975. That the country now exports coal indicates that she is in a position to meet any sudden increase in the home demand of coal.

While self-sufficiency in coal has been achieved, the inadequate crude oil production has led to very heavy dependence on oil imports. Although, crude oil production increased from 2.5 lakh tonnes in 1950-51, to 290 lakh tons in 1934-85, this tremendous increase could not make India self-sufficient in oil. It is because along with increse in production the consumption has also been increasing (31 lakh tonnes in 1950-51 to 388 lakh tonnes in 1984-85). As the demand is more than the supply, India is compelled to import large quantities of crude oil from other countries, specially Iran. In spite of the efforts of Oil and Natural Gas Commission (ONGC), which is engaged in the work of exploration and exploitation of oil, the supply has fallen short of the growing demand. The production meets two-thirds of the domestic demand. The government has- tried hard to bridge the oil gap and succeeded in reducing it from 147 million tons in 1975 to only 98 million tons in 1984-85. Nevertheless, making the country self-sufficient in oil appears to be a distant possibility. Heavy dependence for this strategic fuel on foreign sources is considered to be a serious weakness of India.

The most convenient and versatile form of energy, the water power is next to coal as supplier of energy. Its demand has been growing at a rate faster than other forms of energy and as an inexhaustible source of power, electricity plays a crucial role in both agriculture, industry and transport. But hydroelectricity is in a backward stage of development in India. By 1985 only about 11.0% of the hydroelectric potential of 89.2 thousand MW had been developed. The installed capacity of hydroelectric power plants was 19.0 lakh kw in 1947, which rose to 114.0 lakh kw at the end of the Fourth Plan and to 155.0 lakh kw in 1985-86. The installed capacity has been increasing at a fast rate as a result of the construction of a number of major river valley projects, like the Damodar Valley cooperation (DVC), the Bhakra Nangal, Hirakud etc. Inspite of the phenomenal growth of Hydel power, its share in the total production of 170 million units of electricity in 1985-86 was only 30% (Coal was still the major source of electricity and contributed 67% of the total production. The nuclear plants contributed only 3%).

India is a peace loving country and is committed to the constructive uses of nuclear power, which has been developed on a modest scale. At the end of 1985-86 the, installed capacity of nuclear power plants was 12.7 lakh kw. The four neclear power stations are Tarapur near Bombay (Maharashtra), Rana Pratap Sagar (Rajasthan), Kalpakkam (Tamil Nadu) and Narora (U.P.). As mentioned above, nuclear power constitutes only 3% of the total generation of electricity in the country. It is heartening to note that the country has one of the world's richest deposits of thorium about 2.0 lakh tons, with 9% concentrate.

MINERALS AND OTHER RESOURCES

Much has been written on minerals and world politics in political geography and cognate disciplines. I here is no denying the fact that minerals have influenced international relations in all part of the world. History is replete with examples of interstate rivalry over access to mineralized areas, "Mineral resources may mean wealth specially to a small nation; however this situation may be temporary and may involve the danger of foreign control or interference". Since minerals provide one of the bases of industrial strength, they are much in demand in the world today.

No developed nation is entirely self-sufficient in minerals. 'A country may be sufficient in one or two minerals. But there may be some important minerals which need to be imported from other countries because of inadequate home production. Still there are minerals for whose supply the nation depends wholly on foreign sources. While procuring them may be easy during peace time, their supply may be completely stopped during the war. For example, the U.S.A. produces sufficient coal and iron ore. But for minerals such as mica, manganess, tin, and chromium she depends entirely on imports. As the lack of these essential minerals during the war inevitably proves to be a serious handicap, they rank high on the country's list of strategic minerals.

A striking feature of mineral resources is that they are distributed on the surface of the each much less regularly than cultivable soil. Nature seems to have been rather whimsical in scattering them through the earth's crust irregularly. This has given rise to international trade in minerals on a large scale. Another significant feature is that the mineral resources are exhaustible and incapable of being renewed quickly. Each deposit of a mineral if worked continuously will ultimately be over at any point of time. As a factor of state power minerals therefore, provide a great amount of uncertainty regarding their future availability.

India's Mineral Resource Base

Fortunately India is richly endowed with minerals, specially iron ore, manganese, mica, bauxite, limestone, magnesite, titanium and thorium. There are a few strategically important minerals which India does not possess at all or in sufficient quantity. They are' antimony, cobalt, mercury, molybdenum, nickel, tin, tungusten and vanadium.

Mining has recorded great progress in recent years. The value of mineral production in India (excluding atomic minerals) has increased from Rs. 1,112 crores in 1975 to Rs. 8487.3 crores in 1985. The share of fuel minerals (coal and oil) was however the largest being 87% of the total value, followed by non-metallic minerals 7% and metallic minerals 6%. The index of mineral production during 1985 (base 1970 = 100) was 240 as compared to 129 in 1975. The metallic minerals which India produces are iron ore, copper, chromite, manganese, lead, zinc and bauxite. India's non-metallic .minerals include limestone, dolomite, gypsum, apatite and phosphorite, mica and magnesite, Among the precious minerals gold and diamond are worth mentioning. India is now a significant exporter of iron ore, mica and manganese.

Resources other than minerals include agricultural fibres like cotton and jute, fibres such as wool and silk, and forest products nemely rubber, timber and wood pulp. These resources are important economic elements of political geography since they form a significant basis for the industrial development of a state.

INDUSTRY

"A highly developed manufacturing industry is the most conspicuous determinant of power. Every great power of modem times has been an industrial power". The distinction is often made between a developed country which is highly industrialized and a developing country which is not. As a matter of fact political power is synonymous with industrial power. All politically great nations have made phenomenal progress in mañufacturing. Industrialization generates economic strength and is therefore considered to be a key to political power. The German Empire before the First World War was built on coal and iron. "The skilled exploitation of the great coal fields of Ruhr, upper Silesia and the Saar alone made possible the development of steel, chemical and electrical industries which established her as the first industrial nation of continental Europe".

Industrial capacity has been considered to be a good measure of national power. Several methods of measuring industrial capacity have been evolved. The amount of industrial production, specially of steel production, gives one such measure of the industrial strength of country. Another method of assessing industrial strength uses the percentage figure of industrial workers to the total population of the country. It may be inferred that higher the industrial production or the percentage of industrial workers to total population, greater is the industrial strength of the state. It seems that there is a direct relationship between industrial production or percentage of industrial workers and political power of the state.

Doubts have, however, been expressed as to the efficacy of these methods. The absolute or percentage figure of population employed in industries is not of much value since the effectiveness of the industrial labour varies from country to country because of unequal technological development. Moreover, uniform statistics are not available and some states even do not publish such statistics. Pounds has remarked that industrial production is no better index of political power for "it is difficult if not impossible to equate steel with textiles or chemicals with consumer goods... Furthermore, within each of the main branches of industry, how should one distinguish between producer of low grade steel such as prance and the producer of high-grade and alloy steel such as Sweden; between the chemical industry which produces fertilizers and that which turns out high explosives or plasticts or drugs. It is (therefore) doubtful that a rating of state on any absolute basis is possible".

India's Industrial Strength

India has emerged as one of the important industrial nations of the world on account of her most spectacular and impressive performance in manufacturing industries since independence. The industrial production has been showing an upward trend during the period of planned economic development. While the general index of industrial production was 109.2 in 1961 (base 1960 = 100), it rose to 184.3 in 1970. The trend in industrial production has been rising during the 1980's. The index of industrial production (base 1970s 100) has increased from 154 in 1980-81 to 209 in 1985-86.

Not only has industrial production increased but its share in the net domestic production has also increased from 13.4% in 1970-71 to 15% in 1984-85.

Although, Indian economy is still largely agrarian, its industrial structure has grown much stronger since the time of independence. During the five year plans Indian industry has not only been expanded, but also greatly diversified. The expansion and diversification have been achieved by establishing new units of the existing industries and setting up several new industrial enterprises. For instance, India had only two major iron and steel units in the beginning of the plans. Now there are six major steel plants with an annual capacity of about 89 lakh tons. At the same time several new industries were set up. These are : electronics, engineering, tractor, locomotive, automobile, communication and power generation equipment, chemical fertilizer, plastics and drugs. The textile industry has been diversified by adding rayon and other units.

The expansion of the public sector is another land-mark of the industrial growth of India. Whereas in 1951 there were only five public sector undertakings with an investment of Rs. 29 crores, their number in March 1985 rose to 221 with an investment of Rs. 42811 crores. The public sector undertakings produce a large number of products such as steel, aluminium, copper, heavy and light engineering, fertilizer. Chemicals, drugs, locomotives, electronics, aircraft and ships. Having achieved self-sufficiency, it now exports its products of pig iron, steel, machine tools, rail coaches, industrial boilers, electronic instruments and power station equipments to other countries of the world. At present India's major industries are the metallurgical industries, mechanical and electrical engineering, chemical and allied, textiles and food industries.

TRADE

Economic self-sufficiency is an ideal condition which no state can ever achieve since it cannot produce everything in adequate quantity within its own territory. Goods not produced or produced inadequately have to be imported from other countries. No developed state is completely self-sufficient and free from dependence on imports. A state which produces a commodity at a high cost can import it more cheaply from another country, which has a large surplus for export. Thus, the necessity of international trade arises and all the states of the world are engaged in it.

The international trade gives rise to two problems. Firstly, the state with heavy dependence on import trade is always in a vulnerable position. For example. Great Britain has excessive dependence on food imports and therefore had to face great difficulty during two world wars because supply routes were blocked. Secondly, the state may face difficulty in making payment for the imported goods. The payment is made by means of exports, by means of services performed by the importing country and by means of payment in gold, securities etc. However only a few countries such as the U.S.A. Russia and Japan can pay by means of services. The payment in gold etc. is made only in the time of emergency. It is therefore only by means of exports that payments are mostly made. It is for this reason that states try to promote their export trade.

There is no complete: freedom in international trade which is conducted by agreement reached between the governments. Each state controls the direction and volume of trade. It may refuse the export and import of certain goods. The state may regulate trade by quota system. It may impose tariffs usually on imports and sometimes on exports. But no state would like to lose revenue from duties imposed on trade. Some states, specially powerful ones, try to use their trade policies for fulfilling their political objectives.

The underdeveloped states are primary producers of raw materials from the forest, field and mine, while the developed countries import the raw materials and after converting them into manufactured products export them to industrially backward states. The developed states use their trade policy for political ends by pre-emptive buying and preferential tariffs. "Preemptive buying, in short, is the buying up of all or most of an available produce in order to prevent any one else from doing so.

This may be done in order to prevent a strategic material from falling into the hands of an enemy. This may also be done to place the seller in a dependent position politically". The state increases tariffs on the goods imported from another state for protecting its own industry. It may charge preferential tariff from a friendly state which it specially desires to favour. India has adopted the policy of conducting its trade with Nepal and Bangladesh on a "most favoured nation" basis.

After the Second World War the General Agreement of Tariff and Trade (GATT) was reached at Geneva in 1947. The agreement framed rules for the smooth conduct of international trade. It has succeeded in removing the artificial obstacles to international trade and the states now have wider freedom of conducting trade than before the GATT was signed.

India's Foreign Trade

Since 1950-51 the value of both imports and exports has been rising. The most striking feature of the foreign trade of India is that the balance of trade has always been, unfavourable signifying the excess of imports over the exports. India's exports during 1984-85 amounted to Rs. 16,855 crore and imports to Rs. 17,173 crores, leaving a trade deficit of Rs. 5,318 crore. In 1988-89 India established an all-time record of Rs. 20,281 crore in its export trade. But as the imports amounted to Rs. 27,693 crore, the trade deficit further rose to Rs. 7,412 crore. This is not a very happy situation. The ideal situation is one in which the imports and exports are almost equal. But it is very difficult to achieve that ideal. The national trade policy, announced in April 1985, therefore, aimed at boosting exports and encouraging effective substitution of imports to reduce the trade deficit.

Pattern of Export-Import Trade

Whereas at the time of independence, India exported raw materials and imported manufactured goods, she is now an important exporter of manufactured goods. The exports have not only increased in value but have also been diversified. The important items of exports in order of decreasing value (provisional) in 1985-86 were gems and jewellery, readymade garments, tea, machinery and transport equipment, iron ore, leather and, leather manufactures, fish, cotton, chemicals, jute and spices. The chief items of import trade in 1985-86 were petroleum and petroleum products, machinery and transport equipment, iron and steel, pearls, precious and semi-precious stones, fertilizers, chemicals, vegetable oils, non-ferrous metals, artificial resins, plastic materials, paper and paper board. For the speedy growth of Indian economy imports of petroleum and its products, specific type of machinery spares and components, iron and steel, fertilizer, non-ferrous metals and other industrial raw materials are required and India will continue their import for the time being. Significantly the first seven items of imports make up over 70% of the total imports.

Direction of Foreign Trade

Although, India has trade ties with almost all developing and developed countries of the world, three major trading partners are the U.S.A. Russia and Japan. The dominance of the U.S.A. in both import and export trade is most striking indeed. Besides the above three, other important countries with which India has trade relations are Germany, U.K., Belgium, France, Italy, the Netherlands and other E.C.M. countries, Saudi Arabia, U.A.E., Iraq and Kuwait in West Asia and Iran, Singapore, Hongkong and Australia of ESCAP (Economic and Social Commission for Asia and Pacific). One striking feature of the foreign trade of India is the unfavourable balance of trade with almost all legions of the world. Hence, it is necessary to increase the value of exports so that the trade deficit which at present characterizes India's foreign trade is wiped out expeditiously.

TRANSPORTATION

"Transportation is a major factor in the evaluation of nation in peace as well as in war". There are two aspects of transportation. One is external and is concerned with the international trade which is by land (rail and road), water (ocean, river, lakes and canals) and air to a very limited extent. The other is internal which plays a very important role in the economic and political development of the state. Transportation is significant in national defence, political unity and economic development of the state. Thus an efficient means of transportation of men and material from one place to another is a necessary element of the national power.

In war, defence of a country is possible with the facility of transporting army personnel and military equipment from one part of the country to another. The strategic roads and railways are therefore, necessary to strengthen military power of a nation.

The unity and strength of a nation depend to a large extent on the lines of transport and communication. Many states of the world possess a large network of roads and railways radiating from the national capital to all parts as in India, France and Russia. A well developed transport system is considered necessary for the political unity of the state. New Delhi is very well 'connected with all parts of India by roads, railways and airways. As a matter of fact the unity of British India was a creation of a large railway network. "Before the communist Revolution, a rail road was built from Moscow to the Ural Mountains and across Siberia to Vladivostok. Since then this line has been double tracked, supplemented by a net of branch lines and made more effective". The trans-continental rail lines of Canada and the USA have played a major role in uniting the eastern parts with the western sections of each of the two countries.

Without an efficient network of transportation lines raw materials from the field and mines cannot be assembled at the factory sites and similarly manufactured-goods cannot be brought to the market places. The economic development of a country cannot be achieved with a faster rate without an efficient network of transport. During drought and famine food supplies can be transported to the affected areas in no time.

The railways are supplemented by roads and inland waterways in an overall transport system of a country. Roads can be built in the mountainous areas where rail lines are difficult to construct. The water transport is relatively cheap and suited to the movement of bulky goods. It is possible in all countries where there are navigable rivers, lakes and canals. India has an ambitious plan to develop the Ganga river as an inland waterway from Haldia near Calcutta to Allahabad. In European countries such as France, Germany and Russia a large number of canals have been cut to interlink the navigable rivers to make inland water transport more effective. The Great Lakes of North America are the finest example of transportation by lakes.

Transportation in India

There is no exaggeration to say that railways are the lifelines of India. They are no doubt the chief means of transportation with a total route kilometerage of 61,850 (March, 1985), out of which 6,325 km are electrified. The Indian railways are the largest in Asia and fourth largest in the world after the U.S.A., Russia and Canada. For administrative purposes railways have been divided into nine zones : Central, Eastern, Northern, North-Eastern, North East Frontier, Southern, South Central, South Eastern and Western, It is needless to say that railways have played a major role in national unity, defence and development of India.

The road network of India is one of the largest in the world. The total length of roads, both surfaced and unsurfaced, was 15,54,204 km, in 1982-83. Indian roads are classified into national highways, state highways and rural roads including village and district roads. "Roads are being developed in rural areas under the minimum needs programme, the objective being to link all villages with a population of 1500 and above and 50 percent of the villages with a population of 1000-1500 with all weather roads by 1990" (India 1986). The roads have strengthened military defences considerably in the remote and inaccessible areas. "The Border Roads Development Board was setup in 1960 for accelerating economic development and strengthening defence pre-paredness through rapid and coordinated improvement of road communication in the north and north-eastern border areas".

In India river transport was much developed before the coming of the railway. As a matter of fact, the decline in river traffic was caused by the use of railways. Further, the irregular water depth and shifting channels due to the seasonal character of rainfall and diversion of river water through canals for irrigation led to the decline of river navigation. Although, the country has 5,200 km. of navigable stretch of major rivers, only 1700 km, are being used. The important navigable rivers are the Ganga, the Brahmaputra, the Godavari, the Krishna, the Mahanadi, the Narmada and the Tapti. The Allahabad-Haldia stretch of the Ganga-Bhagirathi-Hooghly river system will be developed under the National Waterways Act of 1982. The navigation has been started in the Haldia-Farakka section. But river traffic upstream of Farakka will be possible on the commissioning of navigation locks at Farakka. When completed, the Haldia-Allahabad inland waterway will be the longest in India.

The canal navigation is not very popular in India. The length of canals in which navigation is possible is 4300 km. But only 485 km. is suitable for mechanized craft (331 km. are actually in use). The Buckingham canal in the coastal area of Andhra Pradesh is the most famous of all the canals. The

Badagana-Mahe and Cochin-Quilon canals connect the lagoons on the sea coast 'of Kerala.

The air transport in India is operated by two public sector corporations named Indian Airlines and Air India, which were established in 1953. While the former operates air services within the country between New Delhi and all the State capitals and other important towns and cities, the latter operates services to some fifty-nine countries of the world. Vayudoot, the third airline was incorporated as a Private Ltd. Company in January 1981 "to connect inaccessible areas of N.E. region and stations which are not covered by Indian Airlines and are important centres of trade, commerce or are important from the point of view of tourism".

The last but not the least are the pipelines which transport both crude and refined petroleum and natural gas over long distances. In India pipelines have been laid in Assam, Bihar and U.P. to facilitate domestic transport of oil. Of late natural gas is being transported from the gas fields to the fertilizer factories. The Hajira-Bijaipur-Jagdishpur (HBJ) gas pipe line (1730 km.) built by the Gas Authority of India Limited (GAIL) carries gas from Hajira in Gujarat to the gas based fertilizer plants of M.P., Rajasthan and U.P.

THE STATE AS A POLITICO-TERRITORIAL PHENOMENON

"The state is a geographic feature which We visualize on the political map" (Hartshorne). The celebrated American geographer, Whittlesey has remarked. "The Kernel of political geography is the political area." Since the state is the centre of all politico-geographical studies political geography has rightly been described as the geography of states. The definition of the term state is, therefore, necessary. In Political Science, International Law and International Relations, the term state signifies, of course for political purposes, the highest institution in A society confined within a specified area.

The state is as organisation of people and space on the political level. Every state is "one part humanity and one part land" (Ratzel). No state can exist without territory and no territory can be formed into a state without people. Thus territory and people are the two pillars of the entire structure known as state. As Goblet has observed, the state is a politico-geographical complex—one in which, of necessity, there is a territorial and a human element. It-is man-created—the products of joint enterprise between man regarded as political animal and the earth regarded as the political soil.

The three essential components of a state are :

(i) permanent population,

(ii) defined territory, and

(iii) a government capable of enforcing its laws within the state's geographical area and entering into relations with outside states. These three are closely welded together, as would be clear from the following definition: "The state is an area organized politically in an effective manner by an indigenous people with a government in effective control of the area" (Pounds).

The Nation

Although, the term nation is vague and difficult to define, yet it may be taken to mean a group of people, originally heterogenous, but possessing a feeling of unity forged by the age-old traditions and customs and common pursuance of a social and economic life.

Ideally the people of a nationality should speak a common language and follow the same religion. But it is not necessary that people belonging to a nation should have a common language or religion. The sovereign state of Canada has two official languages, that of Switzerland, four, yet there are a Canadian nation and a Swiss nation. This would suggest that a common language is not essential for .national homogeneity. Thus, although language is frequently regarded as the main symbol of national unity, people may speak several languages and yet form a national group.

Similarly, religion is now no more an important factor in national unity. On the other hand, common customs and traditions, folk songs, feasts and festivals and food found in a region, may indicate the existence of one common culture and hence of one nation. For instance, inspite of the existence of several regional languages and religions, 840 million people of India, through centuries of living together, have acquired common historical traditions and are, therefore, united as one single nation.

THE STATE AND THE NATION DIFFERENTIATED

It would be worth while to understand clearly the connotations of the two terms—state and nation. Very often, these two are taken to be synonymous terms. But toe term nation is not an alternative expression for state. The statement, "every state is a nation and every nation is a state", is rather an incorrect statement, for the limits of the nation and the state are, in several cases, not coterminous. The distinction between them is perhaps well brought out by thc statement that state refers to territory, the nation to people. Although, the boundaries of the state and the nation often coincide, as is found in the nation states of Europe, a state may have several nations or nationalities, enjoying a certain amount of cultural freedom. The former U.S.S.R. may be cited as an example "Few states of the world are entirely uninational and the multi- national character of many states clearly shows that state and nation are not identical.

While the state is a political community, the nation is a cultural entity. A nationality may exist independently of a state's political organisation and may spread beyond its territorial limits. If it is so, the state may try to extend its geographical area so as to include all people of the same nationality. Conversely, if a state includes people of a different nationality, particularly near the borders, minority problem may arise and the territorial limits of the state may be altered if the sub-nation is granted political independence. On the other hand, if a state, corresponds closely with the nation. It is most stable and permanent. In such a case, the state is the political expression of the nation and we have what is called as the nation state.

In conclusion, it may be said that the state and the nation are not the same and wherever these do not coincide, territorial problems arise which sometimes baffle all solutions. The so-called morphological approach to political geography was most popular during the chorological phase in the development of geography. Whittlesey emphatically declared that "the kernel of political geography is the political area"; and added that "every political unit describes an areal pattern of nuclear core, constituent administrative regions, problem areas, vulnerable zones, capitals, strategic spots, and boundaries—all affecting its success even if not vital to its persistence" (Whittlesey, 1939; reprinted in Kasperson and Minghi, 1969, p. 29). Likewise, Hartshorne (1968) advocated the morphological study of State-areas as the chief paradigm for politico-geographical studies. However, 15 years later, in his presidential address to the Association of American Geographers, Hartshorne showed his disenchantment with the morphological approach. He wrote:

> this approach calls for a descriptive and interpretive study of the external and internal structure of the State-area, considered as a geographic object. The items are familiar: externally—size, shape, location, and boundaries; internally, the regional breakdown in natural regions, cultural regions, of different kinds of people, and the location of the capital... This is almost pure morphology, and therefore static and dull... It does represent a method or system, but not one that seems likely to produce significant results (Hartshorne, 1950, p. 99).

By 1964, however, Hartshorne had substantially revised his opinion. He regretted having given but scant attention to this topic in his previous studies (1950, 1954, and 1960) and admitted that

> experience in teaching the subject has convinced me that it forms as essential part of the geographical study of any political unit, but... the analysis of morphology of political area must be related to the function of the area, and to the purpose of its organization as a political unit (Hartshorne 1968, p. 27; *this paper was originally presented in 1964*).

The aspects of the State as a spatial entity, therefore, deserve greater attention than they have received lately.

Each State represents a somewhat unique structure with respect to the nature (*i.e.*, the physical attributes) of its territory, the changes brought about in the landscape through the process of man's occupance of the territory, as well as with regard to its relationship with other States. The chief elements in the structure of any State are:

(i) An expanse of territory delimited by boundaries,

(ii) An effective' administrative system which binds the diverse regions of the political area together as interdependent parts of a larger functioning whole,

(iii) A resident population,

(iv) An economic structure, and

(v) A circulatory system. These are the fundamental elements in the spatial structure of every State, even though States vary greatly in respect of each of these elements in detail.

AN EXPANSE OF TERRITORY DELIMITED BY BOUNDARIES

Every State occupies and controls a certain well-defined portion of the earth's surface over which it exercises complete sovereignty. The State is, therefore, essentially a territorial phenomenon. No State can exist without a territory, the home base on which the people constituting the nation live, derive their livelihood from and with which, in course of their long historical association, they often develop such a close identity that the territory is transformed into a motherland (or fatherland) for whose safety, security and "honour", they are prepared to make the greatest sacrifices. To clearly define the outer limits of its territory, each State requires well-defined boundary lines demarcating its territory from those of its neighbours.

The word "territory" implies not only the areal expanse of the State, but it also includes the physical attributes of the State's geographical area.

Physical Attributes of Territory

This includes factors like location, size, and shape of State areas as well as their climate, surface configuration, soils, flora and fauna, water features and underground resources, especially minerals, and resources pertaining to food, industrial raw materials, and energy. These physical attributes, each by itself as well as all of them together, place constraints (even without determining their nature in any absolute sense) on the types of activities that the citizens may carry on within the State. But within the broad limits imposed by the diverse physical attributes, the people in each State develop their territory in accordance with their own predilections born out of their historical traditions, cultural traits, and the level of scientific and technological advance.

Location

This is, in many respects, the most important single factor in the physical make-up of a State. Location is a very comprehensive term which includes location with respect to latitude and longitude, land and water bodies, as well as location in relation to other States, particularly the more powerful ones in contemporary international politics. Two other aspects of location are: location in terms of accessibility (*i.e.*, central vs. peripheral location) and strategic location.

Astronomical Location : Astronomical location tells us where a particular region lies on the globe. It describes the location of a place in terms of its latitudes and longitudes. Astronomical location, acting indirectly through climate, exerts a powerful influence on the economic structure of a country and often plays a fundamental role in the country's relationship with other States of the international system. Climatic characteristics determine, to a great extent, the geographical location of the most "favourable areas for human development at the given stage of technological advance. It is in the zones of highest efficiency that the leading powers of any given period in history were located. Today we see that all the great powers are situated largely in the temperate zone. This, however, was not always so. Early civilizations grew up in subtropical and tropical desert regions. There were several reasons for this—deserts surrounding the river valleys provided the much needed protective barrier for unhampered growth of civilization. In the early State of material technology, it was necessary for people to live within a climatic environment that presented no serious problems in regard to shelter and warmth. Only when the level of scientific knowledge had advanced further, and man's ability to cope with harsher environment had considerably increased, did the centre of gravity in the development of civilization shift from the tropical to the temperate regions of the middle latitudes.

> It may be; however, that even as improvement in technology and science permitted advance into the more rigorous areas of the middle latitudes, so further progress in scientific and technical knowledge may make the tropical and polar lands more usable and habitable (Carlson, 1971, p. 26).

Location with respect to Land and Water Bodies : One of the most persistent concepts of political geography has been concerned with the location in relation to the open seas. This was the fundamental thought behind the Heartland theory which may be seen as a special, period-bound, example of the politico-geographical interpretation of continental and oceanic relationships, tested by history, time and again. The general belief that landlocked location poses serious disadvantages to the concerned State, is correct in many respects; but in a strictly strategic sense, a landlocked position may provide a nation engaged in war with the advantage of "inner line".

> Given a system of good communication, a well-developed system of intelligence, and good armies under able leadership, a country can use its interior location to shift troops from one front to another and thus win victories by local superiority. Frederick of Prussia, Napoleon, and also Bolsheviks during the Civil War of the Revolution made the best use of location factors of interior location which except for the advantages they offered in war strategy, were highly disadvantageous (Eric Fischer in: Weigert et al., 1957, p. 175).

The disadvantages of continental or interior location are, however, many and far-reaching. A landlocked country occupies an isolated position with regard to international trade and commerce since such a country is deprived of the opportunities for direct contact with countries with which it does got share common boundaries. This inhibiting factor in the life of continental countries still persists even though the advantages which the maritime countries possessed over those having continental locations before the advent of mechanized land transport (through railways and the automobile) and air carriers, are no longer as pronounced as they were until the beginning of the present century. However,

> even though the process of technology has aided greatly in the utilization of diversified land areas and in establishing continuity and compactness of the territory, the fact has not been altered that every country remains dependent on one or all of its neighbours. Modem industrialization and modem commerce with their dependence on a great variety of raw materials have rather sharpened this relation (Fischer, 1957, p. 176).

It can be understood, therefore, why landlocked States have been actively trying to acquire free access to the sea. History is full of conflicts between continental States trying to expand towards the sea and the coastal ones trying to block their plan of access. This is generally amplified by the Russian example.

> Much of Russia's history can be understood if we see it as a continuous struggle, kindled time and again by the urge to the sea. In earliest limes this urge was limited to attempts to utilize the navigable rivers. The first time Russia reached the sea it was at the Arctic Ocean and this proved to be of very limited value both because of the inhospitable and remote nature of the sea, and of the long and difficult access to the coast from the interior. Only since World War I has Murmansk been connected by a railroad to the interior. Gradually, several seas were reached: in the southeast in 1557, the Caspian Sea at Astrakhan; in the east, in 1635, the Sea of Okhotsk; in the west, in about 1700, the Baltic Sea at St. Petersbury.... in the south, in 1713, the Sea of Azov and through it, in 1783, the Black Sea. All these exits proved unsatisfactory. Either they led into enclosed seas, such as the Baltic and the Black Sea, or over immense, almost uninhabitable stretches, as to the Pacific Ocean. Although frontages at the seas have expanded, they have remained basically unsatisfactory. The [former] Soviet Union, as the heir of Russia [had]

> established itself as the paramount, hardly challenged power in the Baltic, Black, and Caspian seas, but still the "exits from the first two seas [remained] in foreign hands, and the Caspian Sea had no outlet... The [Soviet] pressure on Turkey to deliver the straits into Soviet hands and the pressure on Scandinavian States and Denmark to open free access to the Atlantic could not be explained by Communist ideology or temporary constellations; they [were] inherent in the disabilities of landlocked position (Weigert et al., 1957, p. 183-84).

It is true that significant progress was made both by the earstwhile Soviet Union and Canada to utilize the Arctic Sea by means of ice-breakers, aviation, and the establishment of a weather station. Still, however, the Arctic is of limited value as a highway. But the strategic value of the Arctic is considerable. Great circle routes across the Arctic provide the shortest connections between the Russia and the strategic American bases in the north.

The relative affluence and the wide-ranging scope of the land and sea resources available to any State with a maritime location, exercise far-reaching influence on their economic orientation and, the course of their political advance. The phenomenal rise of British industry was, it is true, partly due to the great advances in the fields of science and technology. But what in fact made the tiny-sized insular State the most powerful industrial nation during the colonial era, was its supremacy over the navigable sea lanes. The colonized lands provided a broad and diversified base for the supply of raw materials on the one hand, and a vast market for its finished industrial products, on the other.

Location with Respect to Other States : This is variously referred to as "vicinal location", "relative location" or "adjacent location". Usually being neighbours with countries of nearly equal power potential or with countries weaker than the State in question, is a distinct advantage, since under such a condition, there are few chances of adverse pressure on international frontiers so that amicable neighbourly relationships can be maintained. This minimizes wasteful expenditure on defence, and makes the scarce resources of the State available for social and economic development. In contrast, whenever countries disparate in size lie adjacent to each other, vicinity more often means domination of the smaller State by the larger and more powerful neighbour. The example of Poland illustrates this point: lying between Germany in the west, Russia in the east, and the former Austro-Hungarian empire in the south, there had been few years when the Polish nation had remained a sovereign political unit.

Location in Terms of Accessibility : This is also sometimes referred to as central vs. peripheral location. Centrality or accessibility in the global context is defined in terms of land mass, population, and economic activity. A central location tends to imply growth in commercial relations between States. The North Atlantic has, in modern times, been the most "central" of all the oceans. This is because of its location between the two most technologically advanced areas of the world—North-eastern North America and Western Europe—with their highly industrialized and somewhat complementary economies. As a result, there has been a very heavy traffic across the ocean connecting the two regions. This is sharply contrasted with the peripheral position of Australia and New Zealand. Centrality of location fosters exchange of ideas, which is a basic requirement for the development of advanced technology.

> The States of the Western World, actively exchanging the skills among themselves and borrowing from others, achieved a unique progress from a primitive base within a short time. The an of the Greeks, the mathematics of the Arabs, the law of the Romans, the industrial techniques of the British have disseminated throughout the West. In the Orient, while Europeans were still rude, the Chinese [and the Indians]... had alresdy reached a high level of achievement in the arts. There was, for a period, a ready exchange of ideas among the people of the East. But the East closed its doors to Western ideas (Carlson, *op cit.*, p. 33).

The concepts of centrality and peripherality are, however, not geographically fixed concepts. Different locations become central or peripheral according to the changing technology as well as changes in the global pattern of economic and political dominance. In ancient times while the Mediterranean was one of the most central areas, whereas Western Europe, lying on its periphery, and the regions of North America were hardly known to the civilized world in Europe and Asia. Since technological development has a direct bearing upon the strategic significance of location relative to other portions of the globe, constant re-evaluation of location is necessary.

Strategic Location : Strategic location is a location that gives to its holder a decisive advantage militarily, politically or economically over its adversaries. A very high degree of strategic importance is given to narrow oceanic straits on important highways of commerce such as the Strait of Gibraltar, connecting the Atlantic and the Mediterranean seas, or the Turkish straits of Dardenelles and Bosporus connecting the Black Sea with the Mediterranean via the Sea of Marmara. The last two straits control the only direct access of the present Commonwealth of Independent States to open seas—a factor which had endowed them with great politico-strategic importance vis-a-vis the former Soviet Union (Fig. 1.1). Likewise, narrow land bridges through which canals are dug to facilitate direct navigation between adjacent oceanic areas, often assume great strategic advantage. An example of this type is the Suez Canal connecting Port Said on the Mediterranean coast with Suez at the head of the Gulf of Suez-Red Sea, and providing a direct and much shorter route from the Atlantic to the Indian Ocean. Before this canal was dug, oceanic traffic between the two areas had to pass round the continent of Africa via the Cape of Good Hope. Another important example is the Panama Canal which provides a direct connection between the North Atlantic and North Pacific Oceans. Before the Panama Canal was constructed, oceanic connection between the western and the eastern coastal areas of the USA was possible only through the long route circumventing the continent of South America (Fig. 1.2). In the days of overland transport, great strategic importance was attached to narrow mountain passes connecting important routes: it was this factor that had made Switzerland strategically so important in Europe (Dikshit, 1975, pp. 32-34.).

The strategic value of any given location must, however, change with changes in the balance of power and technology:

In the eighteenth century Great Britain got possession of the Falkland Islands in the South Atlantic. They were then of strategic importance because they dominated the sea route around the southern tip of South America, and they played an important part in the destruction of the German fleet of Von Spec in

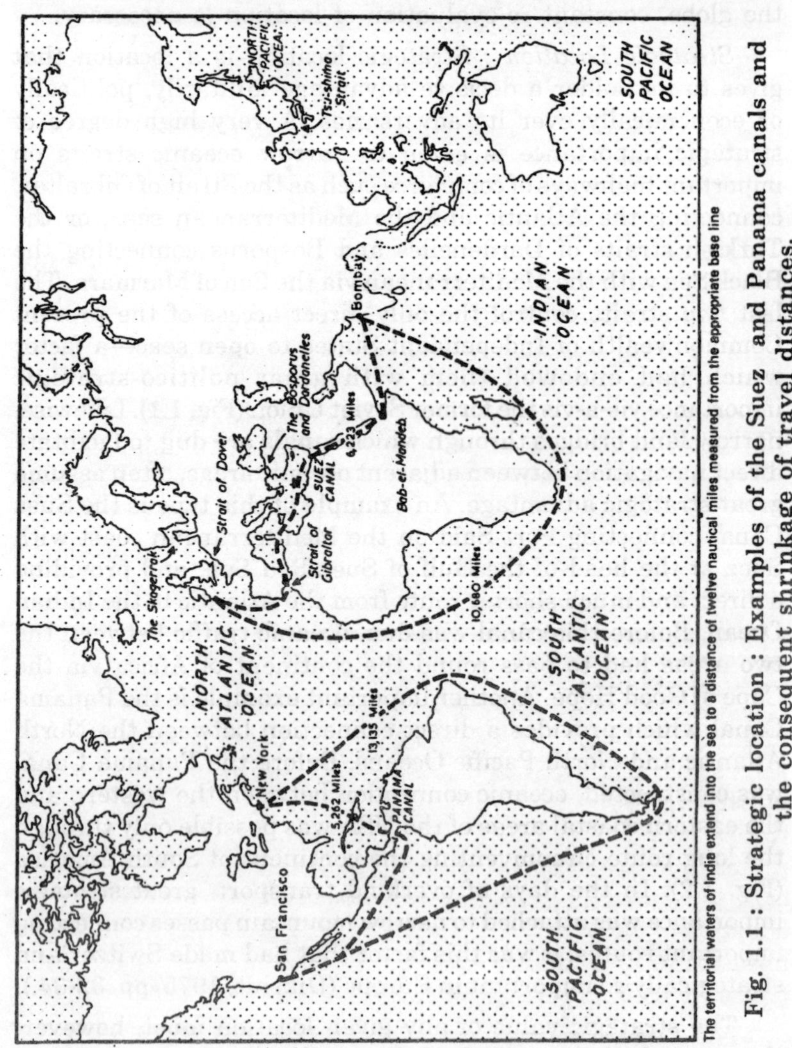

The territorial waters of India extend into the sea to a distance of twelve nautical miles measured from the appropriate base line

Fig. 1.1: Strategic location: Examples of the Suez and Panama canals and the consequent shrinkage of travel distances.

1914. But the opening of the Panama Canal in that year quickly reduced the importance of the Cape Horn route, and now the strategic importance of the Falkland Islands is negligible.

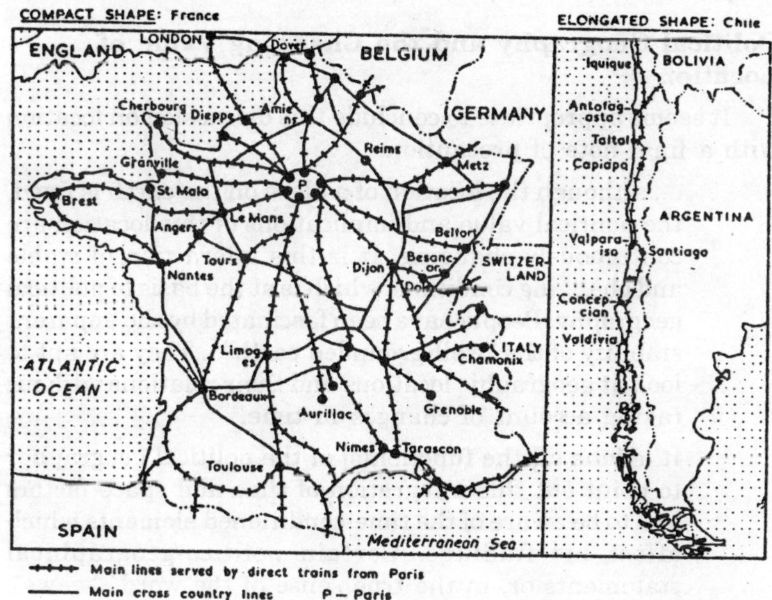

Fig. 1.2 : Politico-geographic value of shape : Examples of France (compact) and Chile (elongated) with their contrasting patterns of transportation networks.

While the importance of the Falkland Islands was diminishing that of Iceland was growing. As long as movement across the Atlantic was primarily by ship, Iceland lay too far north and its seas were too dangerous to be much used by transatlantic shipping. But aircraft flying on the great circle routes from the West European to North American cities, passed close to Iceland and regularly used its airports. Now the general adoption of jet aircraft for long-distance flights has reduced the need to stop for fueling, and in consequence few aircraft now pause at Iceland on transatlantic flights.

The Arctic Ocean, formerly so remote and inaccessible, is now crossed by commercial airlines. And the outer line of

American defence, formerly in Western Europe, the Caribbean, or in the islands of the Pacific Ocean, is now along the DEW (Distant Early Warning) Line in northern Canada and Greenland, and the Arctic wastes of Canada are being exploited for petroleum (Pounds, 1972, p. 64).

Political Geography and the Changing Value of Location

It seems appropriate to conclude this discussion on location with a final note of precaution:

> .... although the location of a place on the earth is fixed, the political value and implications of this location are continuously changing. It is this interaction of stable and changing conditions which is at the basis of political geography. People have been fascinated by the apparent stability of the "well-grounded earth"....They are apt to look at geographic locations and their relations without taking account of changes in time.
>
> It is [one of] the function[s] of the political geographer to point out this integration of time and space factors and to be aware of the time-conditioned elements which affect his findings. Certain politico-geographical statements or, in the true sense of the word, "views", have had great poignancy at one time, but were relevant for a short period only. Others have kept their validity over long periods. Both type of statements are of interest, but should not be confused. Confused thinking on basic concepts of location in political geography, affecting not only the ordinary citizens but statesmen and military strategists alike, is only due to the failure to distinguish properly between the time-bound validity of a politico-geographical concept and its, in many cases only seemingly, timeless application. Such misinterpretation of spatial relationships in location can distort, and has distorted, the outlook of international relations which forms the basis of the foreign policy of great powers (Weigert et al, op. cit., pp. 174–75).

A case in point is the Heartland theory of Mackinder which shall be reviewed in some detail.

As Bradshaw (1971, p. 32, cited in Muir 1975) observed,

> the essence of location approach to the explanation of State behaviour lies in the fact that the geographical location of a State bequeaths upon it certain set of spatial relationships vis-a-vis other states... So many of the potential activities and potential characteristics of a State may be largely attributed to the state's location, that location may become a useful focus for explaining the State's general pattern of behaviour.

It is necessary, therefore, that political geographers should keep adjusting their estimate of the value of specific locations continuously in the light of (i) advances in transport technology, improvements in the systems of transport and the resultant shrinkage in time-distance relationships between places and regions, (ii) the changing global patterns in the distribution of the centres of political and military power; and (iii) decision-makers' perception of the relative location at the given period in history. Human decisions are seldom based on a complete knowledge of reality; more often they ensue from incomplete and sometimes grossly inaccurate perceptions of reality. The task of the political geographer is to analyze the locational perspective of the decision-makers at the relevant time. To view a past decision in any other context or in the context of reality as perceived today, is against one of the fundamental premises of geographical discipline.

Size

Other things being equal, the larger a State is in area, the greater are its chances of supporting a large population and a diversified resource base—the two most important elements in political and economic development.

Positive relationship between large size and politico-economic leadership in the world is largely a modern development. In ancient times, almost all the major States were small in size. Only in small-sized States was it possible for the entire population to be integrated into a dynamic whole that could be harnessed effectively towards the achievement of economic, social, and political power. In those times, large size was a definite disadvantage, because quick means of transport and communication had not yet developed, and the problems involved in administering a large area from a single node of power were stupendous. Even today, a large size has often proved disadvantageous in the relatively underdeveloped economies, where an effective net of transportation, binding the central and peripheral areas through minimization of travel time and cost, has been slow to develop. This is one of the major problems faced by most large-size Afro-Asian and Latin American countries. Among the "developed" countries, the erstwhile Soviet Union presented an extreme case. Here, difficulties arose from the fact that the administrator in Moscow was separated by eleven time-zones from his counterpart in Ulen, on the Bering Strait and was eight time-zones apart from the officials in Vladivostok.

After the French Revolution, the old historical provinces in France were abolished and replaced by more rational administrative divisions (departments). The guiding principle was that the size of each unit should be such that each citizen living in a particular department should be able to visit its headquarters and return home the same day after having attended to his business. The development of modern modes of transport and communications has, however, rendered this original rationale for the small size of the departments as meaningless. After the Second World War when the task of rebuilding the country was being untaken, it was decided to replace the departments by larger administrative units which were more in tune with modern conditions. But before any action could be taken, life had reverted to the accustomed rut, and the old organization continued.

Wherever a positive relationship exists between size and resources (as often is the case), we expect to find an equally positive relationship between size and the minimization of costs in public administration. A large country has to spend relatively less for its centralized services, and owing to its diversified resource-base it has better prospects of achieving self-sufficiency and minimizing dependence on outside areas. There are certain industries which are ordinarily found in larger countries and not in the smaller ones. Larger countries normally possess an automobile industry, an aircraft industry, locomotive building, and heavy machinery, both mechanical and electrical. Smaller countries rarely possess any of these industries.

Large size is of considerable value in security and defence. The larger the area a State controls, the greater are the chances of its political survival in the face of enemy invasion. Only a large country, such as the former Soviet Union during the Second World War, can trade space for time and win, war by retreating hundreds of miles. The time required by the enemy to advance to the retreated position can be used to build ammunition, to train new troops, and to prepare a counter-offensive.

> Only in very large countries can areas still be found which are sufficiently remote from enemy bases to be relatively safe from air attack. Only in a large territory can air raid warning system function efficiently. On the other hand, large size, coupled with other factors, can pose serious problems to military strategy; for instance, outlying parts of a large country, if thinly populated and if lacking in adequate communication lines, are difficult to defend—a problem Russia experienced in the war with Japan in 1904 to 1905 (Weigert et at., op. cit., p. 44).

In respect of internal security, however, large size may often prove a definite disadvantage, since effective communication between the capital and the faraway regions on the borders may be difficult to maintain. This may lead to disaffection and to the rise of fissiparous and separatist movements in the border areas. As Hamilton (1971) noted, awareness of local conditions diminishes with distance from the capital, so that the more remote a region is, the more numerous are the consultations between the regional officials and the decision-makers at the centre. However,

> We must always be aware that evaluations of size factors to military planning and security cannot lead to more than observations of a very general nature. This note of caution has never been more timely than in our age of potential hydrogen bomb warfare. The development of thermonuclear weapons, and the problem of

> radioactive 'fall out', have shattered our concepts of security. In a territory of small size, there will no longer be left any area of refuge or safety. But even for areas of larger size, spatial factors such as remoteness and depth have lost much of their meaning. The horrifying effects of nuclear warfare make it imperative to reappraise the size factor wherever, in yesterday's thinking, it appeared to be an asset to the defence position of a nation (Weigert et al., op. cit., pp. 45-46).

Shape

While very few people would question the political significance of the geographical size of a State, many may .doubt the politico-geographic significance of shape. Pounds (op. cit; p. 54) believed that "geometrical shape of a State presents problems only a degree less acute than those raised by its geographical extent". But many would regard the shape of a Statc's territory as a haphazard characteristic without much political-economic significance. It "seems fair to State that shape in itself has little meaning, but that it has to be taken into consideration as one factor which together with other factors, constitutes [an important element] in the political geography of a country" (Weigert et al., op. cit., p. 73).

Shape may be significant both for defence as well as internal administration. A compact shape minimizes the cost of transportation between the peripheral areas and the seat of government, and thereby, promotes internal unity and coherence. The ideal compact shape is a circle. The closer a State's shape resembles a circle, the greater is the ease in binding the State's centre and periphery into a coherent spatial entity. For internal political cohesion, a circular shape with the capital in the centre, should prove to be of the greatest advantage.

In sharp contrast to this is the situation associated with an elongated shape, which besides adding to difficulties in communication, also increases the length of the boundary. Chile and Norway may be cited as examples:

> Neither has a railroad which runs the whole length of its territory. Norway has an all-weather road, but Chile

does not even have that. Both have to rely heavily on somewhat uncertain coastal shipping, and Chile has been obliged to relax its tariff regulations for its most northerly province and its three most southerly (Pounds, op. cit., p. 55).

However, as Weigert and associates noted,

The claim that a compact and, ideally, a circular shape is the best for the safety of a country is a theoretical deduction without confirmation in experience. The only meaningful question is, whether, and in how far, the political shape parallels certain natural features and factors of human geography. Chile and Norway have an extremely elongated shape, more so than Czechoslovakia, but both have displayed a persistency and stability of shape less subject to changes throughout the centuries than the compact outline of France—Norway, [and] Chile—are largely congruent with a natural geographical region—Where Norway reaches in the southeast across the mountains, it includes all Norwegian-speaking areas. Czechoslovakia did not attain this unity, its physiographic features almost nowhere being congruent with the areas of languages and nations in this area. For this reason and only for this reason, its elongated shape proved so vulnerable (Weigert, et al., op. cit., p. 73).

A compact shape does not always imply cohesion: Romania, one of the most compact States, has always had serious minority problems. Very often a compact shape was achieved at the expense of including a minority group or by excluding a part of one's own people. The western boundary of Romania which was established in 1920 with a view to include a length of railroad, had to include in that process, a substantial dissident Hungarian minority.

AN EFFECTIVE ADMINISTRATIVE SYSTEM

Every State seeks to create a high degree of functional unity and, in certain respects, uniformity, out of the varied geographical area under its direct control. Such functional regions do not occur naturally; only human political agencies, with their effective administrative systems, can create them. As Hartshorne (1968, p. 27) asserted, "considered geographically the purpose of the political organisation of a State is to establish coherent unity and a certain degree of homogeneity over areas which without the State organisation are more or less separate and heterogeneous".

Ideally, a central government should exercise complete political control (in terms of its Constitution) over the entire area falling within the boundaries of the State. This is the test of the effectiveness of a central administration. Few States, however, attain this ideal. In almost every State—except the micro-States with a homogeneous population—there are pockets of dissent and divisiveness. The central administration often endeavours to contain and pacify them by granting special concessions and privileges.

"The form or structure of government by itself, that is, the division of functions among executive, legislative, and judicial branches... is not of focal concern to political geographers", wrote Alexander (1963, p. 51) and added that political geography questions whether there is a central government exercising control over a certain portion of the earth's surface. Associated with this question are considerations pertaining to (i) the type of political control exercised, (ii) the degree of control, (iii) the form of internal organization, and (iv) the effectiveness of the control processes of the central government.

The post-Second World War era is the age of sovereign nation-States, so that the question of the type of political control exercised by one country (like U.K.) over another (like India) which was relevant in the colonial phase of histroy—is no longer so. As such, the nature of political control represented by such terms as colony, protectorate, mandate, trusteeship territory, international zone, and condominium carry little meaning today, and need not detain us.

Degree of Political Control

In theory, a State exercises absolute control over the entire territory within its boundaries. In practice however, the degree of the central government's hold may vary from one part of the State to the other. It is one of the primary tasks of the political geographer to study and analyze this phenomenon. A most important prescriptive statement on this subject was given by Hartshorne's Functional Approach in Political Geography, wherein he recommended that political geographers must concern themselves with the functioning of States as spatial organisations or systems and in that context, they must identify and analyze the forces that tend to bind the diverse regions of the State area together into a functioning whole (*i.e.*, centripetal forces). Also, important are forces that put strain on national unity and create fissiparous tendencies (*i.e.*, centrifugal forces). It was in this context that the valuable concepts of *State-idea and raison d'etre were* developed.

The Form of Internal Organization

For a proper appreciation of the functioning of a State, one must know how its government is organized. The nature of governmental structure has a lot to do with its administrative effectiveness. Broadly speaking, there are two systems of governmental organization—*unitary and federal*.

The chief distinction between the two pertains to differences in the territorial distribution of governmental powers. In a unitary set-up the overall administrative authority is vested in the central government; the local and regional governments (wherever existing) function at the discretion of the central authority. In a federal set up, on the other hand, local governments derive their authority from the regional governments, which themselves possess constitutionally guaranteed autonomy in specified matters of great importance. Since the regional governments derive their autonomous powers from the constitution rather than from the central government, they do not depend on the discretions of the centre.

The Effectiveness of Central Authority

As Whittlesey (1935, pp. 85-97) wrote, political activities leave their impressions upon the landscape just as economic pursuits do. The role of the central government acts as a landscape-modifying agent is an important theme in political geography.

POPULATION FACTORS IN POLITICAL GEOGRAPHY

Although every aspect of population impinges upon politics, one way or the other, the aspects of population of direct concern to the political geographer are: size, distribution, quality, and structure. One of the fundamental factors in modern history has been the phenomenal growth of human population, resulting from the control of diseases and the consequent reduction in mortality rates, accompanied by increases in net reproduction rates. The total population of the world is estimated to have been only about 465 million in 1650; it rose to about 1,098 million by 1850, and stood close to 5,300 million in 1990. Not only has the total world population registered tremendous increase, but there has also been a sharp change in the geographical pattern of its distribution.

Size of Population

Countries of the world vary a great deal in the size of their population. At the one end is the Peoples Republic of China with over 1114 million people in 1990, followed by India with almost 844 million. Next comes the United States with about 250 million. At the other end of the scale are Nauru (8,000). Andorra (51,000). San Marino (23,000). Monaco (30,000). and Iceland (254,000).

As technology advances and civilizations progress, the mere size of population carries little significance. Population must be matched with resources. And it should possess skill in order to be effective. But this was not always the case. The French politico-military supremacy in the eighteenth and the early nineteenth century Europe, was largely due to its large population that supported the largest army in Europe. French supremacy during that period cannot be explained either in terms of internal disunity in the countries opposed to France, or in the enthusiasm that the French soldiers are said to have derived from the new social order created by the French Revolution. It was largely a function of the supremacy of numbers. France had a population of about 24 million at the time of the Napoleonic wars, whereas the Russian total was only about 19 million. All other European powers lay far behind. The total population of Britain, as late as 1801, was only 10.9 million.

The situation changed by the middle of the nineteenth century, when the rate of growth of population in France became one of the lowest in Europe. By 1870, Germany had become more populous than France. Even Great Britain overtook France by the turn of the century. It would be naive to explain the rise of Germany in the latter half of the nineteenth century, and the displacement of France as the leading European power after the 1870-1871, war merely in terms of the changed balance of population, even through "the efficiency of the German army as a fighting machine was not unrelated to its size, and without the changed population ratio, the success of Germany might not have been so clear-cut as in fact it was" (Pounds, op. cit., pp. 123-124).

Modern warfare, since the Second World War, is no longer primarily a war of numbers. Yet, most military conflicts today are limited conflicts, where conventional methods of war are still of primary importance. In all such conflicts, the weight of sheer numbers remains a significant variable as was demonstrated by the American experience in Vietnam. So long as there is a clear possibility that the deadly weapons of large-scale destruction would not be used in a future war, the population factor remains a major consideration in the assessment of the war potentials of States. But, in the present age of global politics, even local conflicts have a tendency to escalate into international conflicts of world-wide involvement, so that the military significance of the sheer size of population is rather limited.

Density and Distribution

Politically and economically, far more significant than the size of the population are its density and the pattern of spatial distribution. Density of population is obtained by dividing the total population of the State by its total area. It is therefore, always described in terms of per unit area. It indicates the extent of the pressure of population on resources. Earlier, when numbers meant power, overcrowding in the case of powerful States (like Germany and Japan) created, an urge (and provided a pretext) for territorial expansion and war.

Figures for the density of population do not, however, reveal much by themselves. They would not tell us whether the population is evenly distributed, creating a uniform pressure on land or whether, as in the case of Egypt, large areas remain desolate and most of the population is concentrated in a small area. Besides, while in some States there may be only one node of high density, in others, there may be several such nodes separated by diversities of culture, tradition, and many other natural or man-made barriers. Generally speaking, the larger the number of these nodes, the greater are the chances of existence of rival sub-nationalities. Hence, greater the strain on the central authority in the task of State-formation and nation-building. The geographical distribution of population is, therefore, an important element in the political geography of any State.

Quality of Population

In the present technologically advanced civilization, the quality of population is far more significant than either size, density or distribution. In an underdeveloped country, with a great mass of illiterate, half-fed and diseased population, a certain density may create a seemingly unredeemable economic and political situation. In a more developed country with a high level of education and technology, and high standards of health and living, the same level of density may present quite a comfortable situation owing to factors like alternative avenues of employment. As a general rule, under-developed countries have the lowest educational levels and low standards of health, making for poor quality work force. In such States one of the main policy objectives is population control, in order to ensure a minimum standard of well-being to all.

The Structure of Population

The structure of population in terms of age, sex, birth rate, death rate, level of technical efficiency and ethnic-cultural diversity are the elements of real significance.

One of the commonest methods of representing population is by means of population pyramids (Fig. 1.3), in which the total population in specified age groups (usually for five-year periods) is represented by horizontal bars arranged one above the other so that the youngest group is at the base and the oldest at the top. These bars are commonly divided by sex. The population pyramid for a country having stable population will contract only gradually upward, but for a country with a decling population through diminishing birth rate the base of the pyramid will be narrow because the population of the young may be smaller than that of adults of working age groups. In contrast, in case of countries with very high birth rate, the base of the pyramid shall be unusually large. The population pyramid indicates the nature of the work force available within a country, both at the present as well as in the foreseeable future. It also indicates the status of health and general life-expectancy.

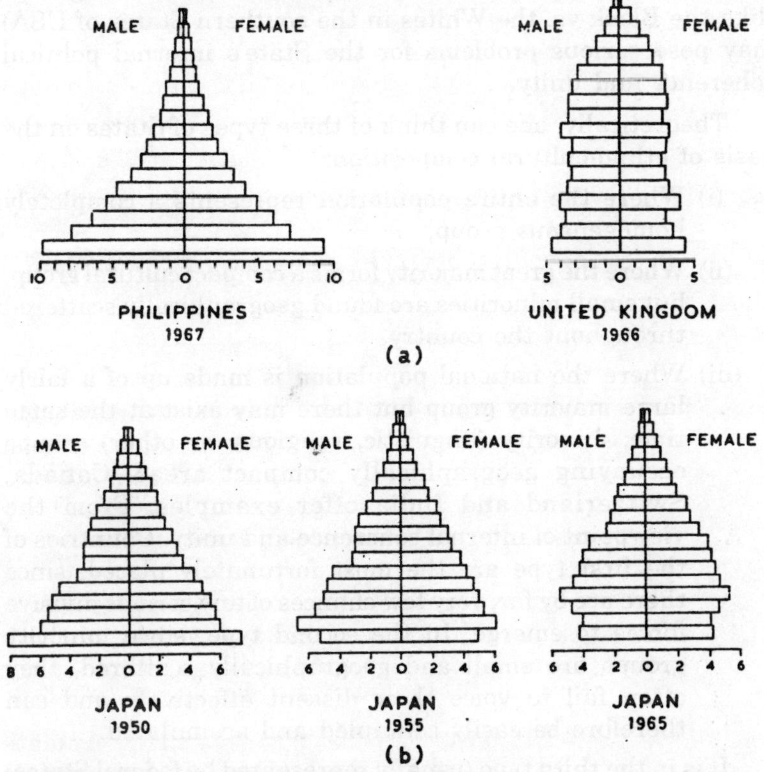

Fig. 1.3 : Population Pyramids : as tools in the analysis of population structure: examples of Philippines, United Kingdom and Japan.

The balance between male and female population (sex ratio) is also very significant. A high percentage of female population may often mean a relatively low effectiveness of population. However in technologically advanced countries, women may stand on par with men in technical efficiency. Sex composition nevertheless remains a significant variable.

Ethnic-cultural Diversity

Another important variable is population composition in terms of language, religion, "race" or "nationality". The presence of sizeable groups of people speaking languages other than that of the majority, or groups following different religions (especially if they had a past history of mutual conflict) or "racial" groups (like the Black vs. the Whites in the southern States of USA) may pose serious problems for the State's internal political coherence and unity.

Theoretically, one can think of three types of States on the basis of ethno-cultural composition:

(i) Where the entire population represents a completely homogeneous group,

(ii) Where the great majority forms a compact cultural group, but small minorities are found geographically scattered throughout the country,

(iii) Where the national population is made up of a fairly large majority group but there may exist at the same time, minority (linguistic, religious, or other) groups occupying geographically compact areas. Canada, Switzerland and India offer examples. From the viewpoint of internal coherence and unity. Countries of the first type are the most fortunately placed, since there are by far, very few chances of any serious divisive forces to emerge. In the second type, since minority groups are small and geographically scattered, they often fail to voice their dissent effectively and can therefore be easily contained and assimilated.

It is in the third type (usually represented by federal States) that most serious problems of divisiveness, threatening political unity, are faced. A comparative survey of various functioning federations shows that a fundamental/factor in federal stability is that the regional identities on which the constituent units of the federation are based should, not be mutually exclusive, even though they may largely be conflicting. When boundary lines of the regionally-identifying diversities in a federation overlap, the cleavages get transformed into linkages and the result is a stable federal partnership. The point is most clearly brought out by the example of Switzerland, where the overlapping boundaries of language and religion have weakened the divisive force of both the elements, for each linguistic group contains representatives of both the faiths, and of course vice versa. Thus, neither side is able to fortify its camp without risking a fair segment of its group in the opposite camp. Thus, neither language nor religion is able to overplay its hand and the fissiparous tendencies have canceled out. When we compare Switzerland with Canada, the moral is easily drawn. Because both the French language and the Roman Catholic faith have been highly identifiable with the province of Quebec in Canada, here the religious and linguistic division, in contrast to the situation in Switzerland, has been continuously plaguing the federal partnership (Dikshit. 1975. pp. 234-235).

The degree of ethno-cultural diversity and divisiveness for any State can be calculated by an index of ethnic differentiation on the basis of the following formula (Piaseeki, 1964, pp. 73-85):

$$S = 100 \sum_{i=1}^{k} n_i^2 / N^2$$

Where n is the size of the ethnic group i, N is the total population of the country, and k is the number of ethnic groups. The lower the resultant index, the higher shall be the degree of ethnic differentiation. The index for complete ethnic homogeneity is 100. This index, however, shows only the degree of ethnic differentiation and not the degree of compatibility or tension between the groups involved. The practical value of the index is therefore, rather limited.

ECONOMIC STRUCTURE

That economic strength is the basic criterion for military and political power hardly needs to be emphasized. It is this factor that has made the United States the strongest country in the world. Economic self-sufficiency is particularly important in times of war. It was owing to their economic self-sufficiency, that the Germans were able to resist the Allied blockades in the two world wars. Since few States can be self-sufficient in all the resources needed for development, the concept of self-sufficiency is largely relative. Not only the economic structure of the State, but its trade relations and alliances also assume major importance in the assessment of its power potential. An assured supply of these materials in time of crisis, is crucial in any war. Modern warfare is greatly dependent on air supremacy and swift movement on land. For both, power is derived from natural oil, which has become a critical factor in world politics, particularly since its occurrence is limited to only certain well-defined areas of the earth. It is for this reason that the oil producing regions of South-West Asia have become politically so important.

CIRCULATORY SYSTEM

When everything has been said about political events and political institutions, there is one fact that underlies all of them: they always involve men interacting over distance. "Ultimately political organization is spatial organization". In terms of sheer position and extent, politics is static and essentially meaningless. It takes on meaning only when it is viewed in its dynamic form. A given

> nation can maintain its existence within its specified physical boundaries only because somewhere within it there is a core, a focus upon which power is centered and from which control over all parts of the nation, even the most outlying, radiates. The dynamics of politics requires that there be effective communication between men occupying various areas of a political entity, and between men controlling different entities. It further requires that there be the possibility of physical movement, that is, of transportation, among these areas and between these entities—transportation of goods, of men, and in times of stress, of means for making war" (Wolfe, 1963, p. 7).

The term circulation is used here in Gottman's (1952) sense to denote "the flow to peoples, armies, material goods, capital, messages, and ideas across the space open to man's activities". It includes transportation as well as communication.

The basic concept underlying any political organization is the concept of authority. In order that a State becomes a strong functional organization, it is necessary that it should possess a strong central authority having effective control over the entire area enclosed within its boundaries. These links foster a sense of belonging together, without which no nation can long remain viable. They are also necessary to effectively meet internal rebellions or external aggression. In order that coherent unity within the State is maintained, it is necessary that the political authority at the centre must be compelling wield authority in the farthest corners of the State. An effective circulatory system is, therefore, of primary importance in the viability of any State.

MODELS OF MORPHOLOGY OF STATE-AREAS

States vary a great deal in their physical attributes and in the pattern of landscape development. For instructional purposes, however, certain models (*i.e.* recurrent patterns) of morphology of State-areas may nevertheless be identified. Whebell (1970) has suggested a three-fold model, summarized below:

The Old World Model

The States of long historical past, are by and large ethnically based. They contain a number of germinal cores that had started independently of one another. Subsequently, as each expanded outward, the intervening spaces were filled in and settled. Owing to this process of development, many States may possess pockets of regionally concentrated diversities around their original cores posing problems for natural unity. Most States in the Old World may represent definite culture areas, so that boundaries separating one State from the others are ethnically determined. Most international boundaries in the Old World are, therefore, subsequent boundaries. In the Old World national capitals represent the traditional seats of power.

The New World Model

In the States of the New World, the present day morphology of State-areas has generally developed without any reference

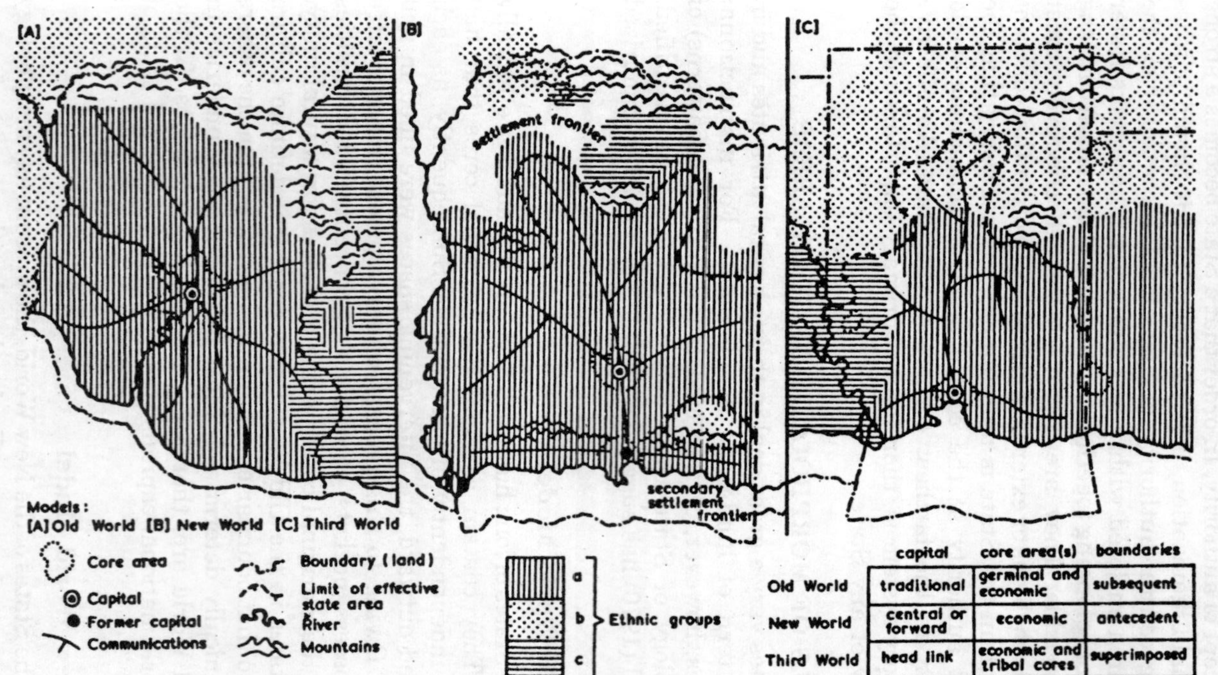

Fig. 1.4 : Models of the Morphology of State Areas : (A) the Old World Model (B) the New World Model and (C) the Third World Model (after C.F.J. Whebell, 1970.

to ethnic differences. These States were started primarily as economic ventures, so that cultural differences were only of minor significance. There is, therefore, almost a complete lack of coincidence between ethnic divisions and international boundaries. International boundaries are antecedent in nature. In most cases boundary lines were decided upon even before the process of settlement had started. Most political boundaries are geometrical boundaries drawn through the vast empty spaces separating the original coastal starting points which often times stand out as economic cores in the present day landscape. While better quality land is occupied by the immigrant population, the original inhabitants have been pushed into extra-ecumenical territory. National capitals are either coastal metropolises of the colonial times or they may be compromise capitals as in the USA, Canada, Australia, or Brazil (Fig. 1.4).

The Third World Model

It represents the case of the post-colonial countries of Asia and Africa which stand almost midway between the Old and the New World models, in the sense that their cultural-historical developments (at least in the cases of countries with continuous historical development overtime) resemble the process of development in the old World. Because of this, they may possess germinal cores, the intervening spaces in between which have long been filled and settled, but pockets of regionally grouped ethnic diversities remain.

Unlike the case of European countries of the Old World, these countries may also possess colonial economic cores that were superimposed upon the traditional pattern. Unlike the European States, international boundaries in the Third World are superimposed colonial boundaries, so that there is no coincidence between ethnic divisions and international boundaries. In many countries in Africa, boundaries divide originally UNITED organic communities. In many cases, the colonial link capitals continue to serve as the present-day capital while in others, old tradition may cause shift of capital to the interior to some traditional seat of power, such as a Delhi or Mandalay.

THE DEVELOPMENT OF STATES

The evolution of the state from small beginning to its present form has had a very long and chequered history. Political scientists, historians and geographers have studied the origin and growth of state from their own points of view. But they generally believe in the evolutionary theory of state growth, according to which the state has passed through several stages, of development. During, its long period of growth it assumed several politico-territorial forms.

From Clan Territory to Tribal Organisation

The transition from clan territory and tribal organization to state community has been a gradual, process. In the ancient period, before the development of agriculture, man led a nomadic life and was of necessity a hunter and gatherer. In the beginning functional, human and territorial units must have been the clan and the clan territory The clan, a large family with a common ancestor, moved about in a certain specified territory. Later several clans forged into a tribe—group of barbarious clans under a recognized chief. Thus clan territory was replaced by the tribal politico-territorial organization. The Romans had originally three (probably representing clans) and ultimately thirty-five such tribal groups. Thus, "first political units came into being when families bonded together into clans in order to defend their claimed territory against intruders" (Norris and Baring). Thereafter socio-political amalgamation of several clans led to tribalism everywhere in the world.

From Tribalism to Civilized State

The change from hunting to pastoralism brought man closer to a defined territory, and he made some advancement from savagery to civilization. Permanent association of human groups to a particular territory was established with the adoption of agriculture as a means of subsistence. The nomadic people began—to lead almost a sedentary mode of life and were fixed to the soil. "Seven thousand years ago along toe banks of the Tigris-Euphrates, Wei, Nile and Ganges rivers, civilizations organized states. The rivers afforded good transportation and the seasonal nature of the climate made necessary a strong government to enforce the growing and storing of food in good years in order to survive the drought that was certain to follow. These ancient river states were the nuclei of the present countries of Iraq, China, Egypt and India." Thus these fertile valleys which supported' dense agricultural populations, due to irrigational facilities and flood control measures, gave rise to more developed organizations of people and territory. Surplus production stimulated political expansion as section of population was enabled to take up specialized occupations in arts, administration, trade and army. Greater commerce led to the development of transport and communication and rise of urban centres. As a result extensive states with large populations came into existence in these few favoured areas of Afro-Asia.

Development of stales in the Mediterranean region. "Many data support the argument that the sequence of events that ultimately led to the spread of the State-idea across the world began in ancient Egypt" (de Blij, 1967). Although civilization in Mesopotamia was the first to emerge in the Middle East, the ancient Egyptian state of the Nile valley came into being much earlier than the one which-rose in the. Tigris-Euphrates valley. The reasons were :

(a) Egypt occupied a pivotal position with reference to Asia, Africa and Europe;
(b) It had the advantage of natural protection; and
(c) Greater surpluses of human and economic resources accelerated the growth of state.

The Egyptian state, though initially completely theocratic, became increasingly militaristic and extended its control as far as Cyprus, Turkey and Syria in the north and Nubea and Cush in the south. Egypt passed through alternate phases of internal decay and consolidation, But it was able to survive for a long time. Egypt's period of true empire lasted from 2650 to 1100 B.C. (de Blij, 1971). Its influence spread far and wide especially in the south-west in sub-Saharan Africa, Where well organized tribal states and empires flourished with great stability and permanency.

Unlike Egypt, Mesopotamia was vulnerable to outside interference as it was the marching ground of different peoples, from the very early times. The land of the Tigris and Euphrates rivers, therefore, witnessed a succession of states. Two empires eventually came into being—Assyria in the north and Babylon in the south. As compared to Babylon, Assyria enjoyed peace and prosperity and made great progress in agriculture and trade. Consequently, as an empire Assyria during the 7th century B.C. expanded greatly and included the whole of Mesopotamia, the Levant and internally weakened and divided Egypt as well.

After the Egyptians, it was the turn of the Assyrians who were conquered by the Persians. The capital of the Persian Empire was Persipolis (now in ruins), located north-east of Shiraz in Southern Iran. At the time of its greatest development, the Persian Empire extended from the Mediterranean Sea in the west to the Indus river in the east and from the Black and Caspian Seas to the Persian Gulf and Arabian Sea. In the Middle East it included Egypt, Syria, Turkey, Mesopotamia, Afghanistan and Baluchistan. The Persian empire, unlike Egypt or Mesopotamia's states, was never highly centralized and therefore did not survive very long and was replaced by Alexander's empire in the fourth century B.C.

While the Middle East witnessed the rise of large states and empires, its immediate vicinity, the east Mediterranean region gave rise to small states. At that time there was one single Greek nation. Therefore a number of so-called city states came into existence. Established by the ancient Greeks, these city states were important landmark is the development of man's politico-territorial organization. Each one of city states possessed a great deal of individuality and was powerful. Another striking feature was that several city states combined to form inter-city leagues for achieving their common economic objectives. Most powerful leagues were the leading cities of Athens, Sparta and Thebes. But no inter-city leagues could last very long due to large scale defection (the last inter-city league, the Acheean, was defeated by the Romans in 147 B.C.). The downfall of the Greek city states was caused by the administrative failure and their inability to control effectively the areas that were captured by military campaigns.

During the fourth century B.C., Alexander the Great marked the culmination of Greek power. He set out to conquer lands lying east-wards and succeeded in establishing a vast empire, which extended from Turkey to Egypt and the Levant to Punjab and Sind in India. But after his death the empire fell. It was now the turn of the Romans to establish their empire in the Middle East.

The Roman Empire

The Roman empire affords an example of the transition of a political territorial organization from clan society, tribalism to a civilized state. In about 400 B.C. Italian peninsula was a territory of many tribes Well fortified hill towns acted as centres of administration and defence, among which Rome was the most important. Gradually the regional influence of Rome increased and by the end of 3rd century B. C. Roman federation extended from the Po valley in the north to the Greek-occupied southern part of Italy. On the unification of the Italian peninsula Rome became a great power in the entire Mediterranean area. Initially the Greeks and Carthagians of north Africa coast put up great resistance. But the Roman empire continued to expand. Ultimately Carthage was defeated and the Greeks admitted Roman authority.

During the 2nd century A.D. Roman empire's greatest expansion took place. It extended upto Britain in Europe and through Greece and Turkey upto Mesopotamia, and in north Africa from Morocco to Egypt. Roman empire thus became a continuous and unified empire strongly centralized on one city, Rome.

Towards the end of 4th century, due to adverse military situation and administrative difficulties the empire broke up into two parts—western empire with Rome as its focus and the eastern empire centred on Byzantine (later renamed as Constantinople and subsequently Istanbul). During the 5th century western empire collapsed but the eastern one as Byzantine empire lasted upto the middle of the 15th century when it was overthrown by the Ottoman Turks.

The last of the great empires to emerge in the Middle East was the Ottoman Empire, centred in Turkey. By the 16th century it was fully developed and included parts of North Africa, S.W. Asia and S.E. Europe. The Ottoman Empire lasted for more than three centuries, and before the World War I, it& power began to show signs of decline and disintegration. It is on the fragments of the Ottoman Empire that modern States of the Middle East have emerged in the present century.

State Formation in the Americas

Before the discovery of North America and arrival of the Europeans, the so-called New World had witnessed the emergence of several indigenous politico-territorial organizations, which were the creations of organized agriculture. Most important of them were the Aztec State of Middle America and Inca Empire of the Central Andes. The Aztec State was "one of the most impressive states to develop anywhere in the world at any time prior to the rise of modern Europe".

But before the rise of the Aztec State, there flourished the Mayan civilization in Yucatan and adjoining area between 300 and 900 A.D. In the early fourteenth century the Aztec state originated when a group of Mexicano speaking Indians founded a settlement on one of the lake islands in the Valley of Mexico. The settlement, named Tenochlitlan, became the capital of a large and powerful state. The Aztec Indians succeeded in establishing their control over the entire valley of Mexico, the heart of the modern, state of Mexico. The State was divided into provinces and districts. There was an elaborate system of tax collection, law enforcement, and rapid transportation and communication. Agricultural productivity was high and irrigation practices were well developed.

Throughout the fourteenth century military strength of the Aztec state was growing and in early 15th century it began the conquest of neighbouring areas. But it could not survive long. Two factors led to the downfall of the Aztec State. The people lived under-constant psychic fear because the Aztec's cult of the sun was based on a large number of human sacrifices.

Secondly, the state had very strained relations with its neighbours. Therefore the Europeans who invaded the state got the full support of the people and succeeded in disintegrating it. Several centuries later the modern state of Mexico emerged on the ruins of a once flourishing, Aztec state.

Another region of indigenous politico-territorial organization was the Central Andes of South America. It was the Inca empire which extended from Ecuador to Central Chile over a distance of 2500 miles. The empire was carved out of a number of elongated basins in the Andes mountains, called 'altiplan' representing mountain valleys filled with eroded material and volcanic debris. The Inca empire was centred upon Cuzco (in Peru), situated at a height of 1100 feet above sea level. Located in the centre of a well watered fertile valley, Cuzco and the surrounding region was the core area of the vast Inca state with far-flung possessions. From this core the Incas began their expansion by military conquest, first to the coastal areas of Peru. Thereafter the empire was extended southwards. At the close of the 14th century the Incas who were good colonizers and administrators succeeded in integrating peoples and regions .of the Andes into a stable and well functioning state.

But the elongated shape of the empire created difficulties in the effective control of the distant regions. The empire faced a number of internal "problems and when the Spaniards invaded it in early 16th century, the empire disintegrated and vanished. It is on the ruins of this vast empire that modern Andean states of Colombia, Ecuador, Peru, Bolivia, Chile and Argentina have emerged on the political map of South America.

Evolution of State in India

According to Sir John Marshall, the acknowledged authority on the Indus Valley civilisation, "India must be recognized along with Persia, Mesopotamia and Egypt as one of the most important areas where civilizing processes were initiated and developed". Six or seven thousand years Old Indus Valley civilization was not only confined to the Punjab and Sind but spread over the Ganga Valley. "Thus it was much more than an Indus Valley civilization" (Nehru, 1960). During the Vedic period northern Indra (valleys of the Indus and the Ganga) consisted of a large number of tribal territories. Historians tell us that during the 6th century B.C. there were sixteen 'mahajanpadas' namely Anga, Magadh, Kashi, Kosala, Vajji, Malla, Chedi, Vatsa, Kuru, Panchal, Matsya, Sursena, Assak and 'Uttarapath'. Except Vajji and Malla, which were 'gantantras' (republics) all others were on 'rajtantras' (kingdoms). "Manu's conception of a state was evidently that of a small kingdom". This conception was however growing and changing, leading to the vast Mauryan empire of the 4th century B.C. and to the international contact with the Greek world.

According to Nehru, two factors led to the emergence of a powerful Mauryan empire. Firstly, there was an urge for building a united centralized state through the amalgamation of petty states and small kingdoms and republics. Secondly, Alexander's invasion of north-west India in 321 B.C. gave the final push to this development. Two men, Chandragupta Maurya and his councillor Chanakya appeared on the scene and took advantage of the prevailing situation. On Alexander's death in 323 B.C. they "raised the cry of nationalism and roused the people against the danger of foreign invasion. This worked and brought help from several hands of small republics and kingdoms. Within two years of Alexander's death Chandragupta Maurya was in possession of Nanda' kingdom of Magadh and with this the Mauryan empire had been established....Except for South India it covered the whole of India from the Arabian Sea to the Bay of Bengal and extended in the north to Kabul. For the first time in recorded history a vast centralized state had risen in India".

The Mauryan empire in course of time faded away and gave place to Sunga dynasty. Later arose the Gupta empire and after several centuries the Muslims arrived in India and established several dynasties and kingdoms, and finally the mighty Moghul empire. On the fall of the Moghul empire, the Britishers succeeded to make this country a part of British empire. On the withdrawal of the British the modern state of India came into existence in 1947.

CAUSES OF RISE AND DECLINE OF STATES

Here it is pertinent to ask what were the causes of the origin of ancient states and empires and why did they disappear from the world scene. Many social scientists including geographers have analysed the rise and decline of States and empires. Professor J.A. Wilson has pointed to three concepts:

1. The principle of challenge and response.
2. The urban revolution concept.

3. Redfield's folk society concept.

The famous historian, Arnold J. Toynbee, introduced the principle of "challenge and response'. Man in a particular situation faced the challenge thrown by nature, for example clearance of forest, growing of crops in the alluvial soils and the control of rainfloods. The challenge demanded human response in the form of improved and even new techniques in various fields of activity. Only when the challenge was sufficiently serious, man sought the solutions. Thus the development process started and great interest developed which must have given rise to states and empires in Egypt, Mesopotamia, Indus Valley and elsewhere in the world. Although, doubts have been expressed about the validity of the principle, it is perhaps the most famous of the three concepts.

The 'urban revolution concept' as defined by Prof. V. Gordon Childe "relates to the way people at a certain stage in history tend to begin to cluster around villages, thus setting in motion a process of agglomeration that eventually marks urban centres out of some of the more favoured agricultural villages" (der Blij, 1971). The surpluses of production and population led to the development of some sort of politico-territorial organization in an urbanized society. The concept may well explain the rise of states and empires around favourably located urban centres. But as in case of the principle of 'challenge and response', Prof. J.A. Wilson has cautioned against the indiscriminate application of the concept.

The Redfield's 'folk society' concept is "another attempt to evaluate the process of urbanization in the evolution of human society". It has been said that it is more likely that folk society gave rise to prehistoric organization, because folk society comparatively, small, homogeneous, deeply religious and strongly traditional has very strong, sense of community and a subsistence economy. Following Wilson, da Blij opines that "the concept is instructive but it does not provide answers to the questions we raised about the rise of states".

Several factors might have contributed to the decay and disappearance of old states and empires. No single factor, however, caused their decay in any one region or no one factor was responsible for the decline of states in different parts of the world. It appears that changes in climatic conditions and economic problems faced by the then governments led to the decline-of extensive empires in the Middle East. Climatic change has been one of the major theories to account for the decline of states, especially in Mesopotamia and North Africa. There are urban sites in the arid regions which have been excavated by archaeologists and it is assumed "that when those towns were thriving centres of government and trade, they must have—enjoyed a moister climate. Afterwards there was a general drying up that was related to the retreat of the warm glaciation in Northern Eurasia.

"The decline of practically every political state—ancient Rome included—can be accounted for on economic grounds". Ancient Egypt was already in the declining state at the time of its conquest by Assyria in the seventh century B. C. It is said that Egypt's prosperity depended on copper which by about 1000 B.C. was exhausted and the Iron Age had started. But there was no iron ore locally available. "Egypt's primacy had coincided with the Bronze Age and it ended when the Iron Age dawned" (de Blij, 1971).

The annihilation of the Aztec state and Inca empire was caused by the invasion of the Spanish people in the early sixteenth century A.D. The indigenous people were conquered and their states were disintegrated to yield place to a large number of small states. Thus in the Americas, factors which led to the decay of states in North Africa and Middle East were not operative.

THEORETICAL APPROACHES TO THE GROWTH OF STATES

Richard Muir has mentioned three distinctive politico-geographical approaches to the question of the growth of states:

(a) Ratzel's organic view,

(b) The core area concept, and

(c) Unified field theory of S.B. Swiss.

In addition, while dealing with the emergence of states in the Middle. East, de Blij (1971), has been another approach in the cycle theory of S. Van Valkenburg. All these four find a reference in this book in different contexts and only a brief description of them is needed here.

Freidrich Ratzel the celebrated German geographer was interested in the relationship of space and state, which he viewed as an organism. He saw a common characteristic of growth m both. But he erred in his analogy, since although organism grows in size, its form is genetically determined before birth, but that of the state is not. In the past, state's spatial growth might have occurred through conquest and annexation but in the present world no state can grow spatially that way. Ratzel's organic theory is, therefore, no more valid. Germans of pre-world war It attempted to expand by waging war and annexing territory of its neighbours but miserably failed.

The core-area concept is more applicable than Ratzel's organic theory to the growth of states. With certain exceptions states do possess core areas in which they originated and later spread from them in almost all directions. These states exhibit a growth pattern which can be fitted with the core area concept. But since all states of the world have not grown from cores, the concept is not universally applicable.

In the unified field theory the emergence of a state is viewed as a product of a series of interconnected stages. An initial idea is developed to produce a political decision which in turn precipitates movement affecting a field within which a political area ultimately materializes. The five stages have been compared to interconnected lake basins (or to hubs of activity) with the dominant direction of flow from idea to area though interaction in the reverse direction is also envisaged : Idea → Decision → Movement → Field → Area

S. B. Jones has applied his theory to the African state of Liberia, Harm J. de Blij to Uganda and Richard Muir to the emergence of Bangladesh. Muir has observed that 'idea' was developed, by discrimination against the people of erstwhile East Pakistan (Bangladesh) by the Pakistan government. The 'decision' was contained in Sheikh Mujibur Rahman's six-point programme for equality and autonomy, military action by Pakistan army, flight of refugees to India; and intervention by the Indian army led to 'movement'; East Pakistan provided the 'field' and independent state of Bangladesh was the 'political area', the last link of the idea—area chain. But Richard Muir is not entirely right in his assessment of the theory when he says that "of the three approaches reviewed" that of Jones appears to be the most universally applicable:" What to say of all, even a majority of stated have not orginated in the manner outlined in Jones' theory.

In the opinion of Blij (1971),' Valkenburg suggested in the cycle theory the possibility of identifying a cyclic process in the political development of nations.' But the cycle theory deals more with stages and types than with the areal growth of states. In the cycle theory described later in this chapter, it is stated that only in the adolescent stage the state is dynamic and aggressive and tries for expansion. The stages of youth, maturity and old age have nothing to do with the spatial growth of a political area. The interruption of the cycle is, however, most striking because it is through it that the state may be brought back to a former stage as it happened in the case of China. The communists disposed of the previous regime of Chiang-kai-shek in 1949 to bring the country back to the state of youth. But it is doubtful whether China, on attaining adolescence, would embarks upon the course of expansion in the present day world.

THE WORLD'S STATE PATTERN

A glance at the political map of the world shows that the entire land surface of the earth is partitioned into a plethora of political units but not all of them are states. Some political areas have yet to emerge as sovereign states. However, it is heartening to note that the post-war period has witnessed a rapid rise in the number of states. Whereas at the dose of the Second World War, the states of the politically divided world numbered 75 only their number rose from 87 in 1950 to 108 in 1958. It stood at 137 in 1968 and in 1970 at 141. In 1976, the number of states was 154, out of which only 145 states were members of the U.N.O. This rapid increase of new independent countries—successor states of colonial empires—has occurred due to the recognition of the right of self determination and spread of intense nationalism and political awakening amongst the peoples of Asia, Africa and other parts of the world. Consequently by 1975, as many as 43 political units in Africa and in Asia attained statehood since the second World War. The process of state-formation has, however, not come to an end. It is expected that the rise in the number of states will continue till the last colonial spot vanishes from the world.

During 1990-93 several changes occurred in the political map of the world. Three European states—Soviet Union, Yugoslavia and Czechoslovakia disintegrated and created 19 new states. Two new-states of Marshall Islands and Micronesia came into existence in the Pacific Ocean. Eritrea in Africa seceded from Ethiopia and became an independent state on May 23, 1993. Two Yemens and two Germanies were united together in 1990.

The emergence of 22 new states (Estonia, Latvia, Lithuania, Kazakhstan, Uzbekistan, Tadjhikstan, Turkmentan, Khirgizstan, Georgia, Armenia, Azerbaijan, Moldavia, Ukraine and Bylorussia—former republics of the U.S.S.R., Slovania Croatia, Bosnia-Herzegovina and Macedonia—former republics of the Yugoslav federation, Slovak republic—eastern Czechoslovakia, Eritrea, Marshall Islands and Micronesia) and the unification of the Peoples Republic of Yemen (Aden) with Yemen Arab Republic (Sana) in May 1990 and of East Germany with West Germany (Bonn) in October 1990 led to a net increase of 20 states from 171 in 1989 to 191 in 1993.

In 1989 there were 171 states in the world—Africa 52, Asia 40, Europe 33, North America 15, South America 20 and Australia 11. At present there are 191 states and their continental break-up is : Africa 53, Asia 46, Europe 46, North America 15, South America 20, and Australia 11. Judging from the number of states Africa is politically the most divided continent in the world closely followed by Asia and Europe with 46 states each.

POLITICO-GEOGRAPHICAL CLASSIFICATION OF STATES

The one hundred and ninety one states of the world are so heterogeneous that they can be classified in broad categories only. Several criteria have been adopted for that purpose, some of which are: evolutionary process, evolutionary stage (behaviour pattern), governmental form, political ideology, extent, location etc.

Drawing the analogy from the agricultural world. Goblet has recognized three types, of states : intensive, extensive and mixed. An intensive state, in its evolution, concentrates on the perfection of its optimum territory. The extensive state merely seeks to extend its domain usually at the expense of other states and the mixed type has an intensive territorial and human core which it often seeks to surround with an extensive territory. A mixed state, thus indulges in purposeful expansion by radiation outward from the intensive core.

S. Van Valkenburg (1939) recognized four stages, namely, young, adolescent, mature and old, based on a cycle in the political development of the state. As there are four-stages, so there are four types of states. Any *young state* immediately after its creation or attainment of sovereignty feels the need of internal, organisation, and devotes all its energy in the consolidation of its internal structure. All the newly independent states of Afro-Asia naturally fall in this group. An adolescent state exhibits the tendency towards the territorial expansion after lift period of internal organisation and consolidation is over. China, Pakistan and Israel may, be cited as examples. A *mature state* abandons the idea of territorial expansion and may even leave certain territory if it does not serve any defensive or economic purpose. Such a state favours international cooperation in world organizations. The U.S.A., U.K. and France and some other West European states, belong to this group. An *old state* is one of declining power and of failing internal organization. Internal disintegration occurs and parts of the state separate and get absorbed in the neighbouring states.

De Blij (1971), however, favours a different nomenclature and says that it is better to classify states according to type rather than stage. Thus, stages may be designed w organizing type (young), expanding type (adolescent), stable type (mature) and disintegrating type (old).

The application of other criteria such as ideology, government and location leads us to a dichotomous classification. According to governmental form states are either republic

(France, India) or kingdoms (U.K., Nepal). Based on federal and unitary forms of government, the states are of two types—federal states (U.S.A., Australia, Canada) and unitary states (France, China, Japan). On the basis of differing political ideology it is possible to make distinction between communist and noncommunist states. Political ideology had divided the modern bipolar world into two power blocs—Eastern and Western. But one should not forget the third bloc of neutral states. If location with respect to seas and oceans is taken into consideration we may distinguish the coastal states from the non-coastal states. Out of a total of 191 states only 41 states are non-coastal and in that way they are unfortunate.

Lastly, the states may be classified on thc basis of their areal extent. The states of the world range in size from the Vatican with an area of less than 0.25 sq. mile to Russia with above 66.7 lakh sq. miles. Geographers have suggested different size categories and tried to put states in one or the other category. Pounds, for example, distinguishes 8 types—giant states (over to lakh sq. km.), outsize states (25 to 60 lakh), very large states (12.5-25 lakh), large states (6.5-12.5 lakh), medium states (2.5-6.5. lakhs), small states (1.25-2.5 lakh), very small states (0.25-1.25 lakhs) and micro states (under 0.25 lakh sq. km. Valkenburg (1963) divides them into 7 categories which are gigantic (over 20 lakh sq. miles) very large (5-20 lakh), large (1-5 lakh), medium (0.25-0.1 lakh), small (0.1-0.25 lakh sq. mile), very small (1000 -10,000 sq. miles), and miniature (less than 1000 sq. miles). Blij's (1967) terminology is as follows: very large (over 10 lakh sq. m.), large 140,000-100,0000), medium (60,000-140,000), small (10,000-69,000), and very small (under 10,000 sq. m.).

Thus, we see that the three geographers have proposed three different classifications on the basis of size. The reader is bewildered to discover that there is no compatibility in them. In Valkenburg's classification a very large state is one whose area is between 5 and 20 lakh square miles whereas in that of de Blij a very large state should spread over 10 lakh sq. miles. Thus for de Blij states covering 5 to 10 lakh sq miles are not very large states. For Pounds only those states whose area is about 5 to 10 lakh sq. miles fall into the category of very large mstates.

2

GEOPOLITICAL WORLD VIEWS : THEORIES OF HEARTLAND AND RIMLAND

INTRODUCTION

Most outstanding British contribution to political geography perhaps came from the geographer-statesman, Halford J. Mackinder (1861-1947). His heartland theory propounded in early twentieth century is still read with interest even today. His earlier works did not arouse as wide an interest in his writings as his famous lecture entitled 'The Geographical Pivot of History', which he delivered before the Royal Geographical Society (Mackinder, 1904). In this paper he advanced the view that the Eurasian core which he named as 'Pivot Area' (later as Heartland in 1919) was inaccessible to sea power and possessed the vast capability of becoming the seat of a great world power which would be able to dominate the whole world.

The Heartland Theory

In 1919 Mackinder published his book 'Democratic Ideals and Reality, in which he formulated his heartland theory. Although Mackinder modified the areal spread of the pivot area in 1919 and again in 1943, the essential features of heartland concept remained unchanged throughout. His theory isolated in three lines :

"Who rules East Europe, commands the Heartland
Who rules the Heartland, commands the World Island,
Who rules the World Island, commands the World".

World's Geopolitical Organization

Mackinder conceived of one landmass which he called the *world island*, comprising of Europe, Asia and Africa in the midst of one World Ocean.

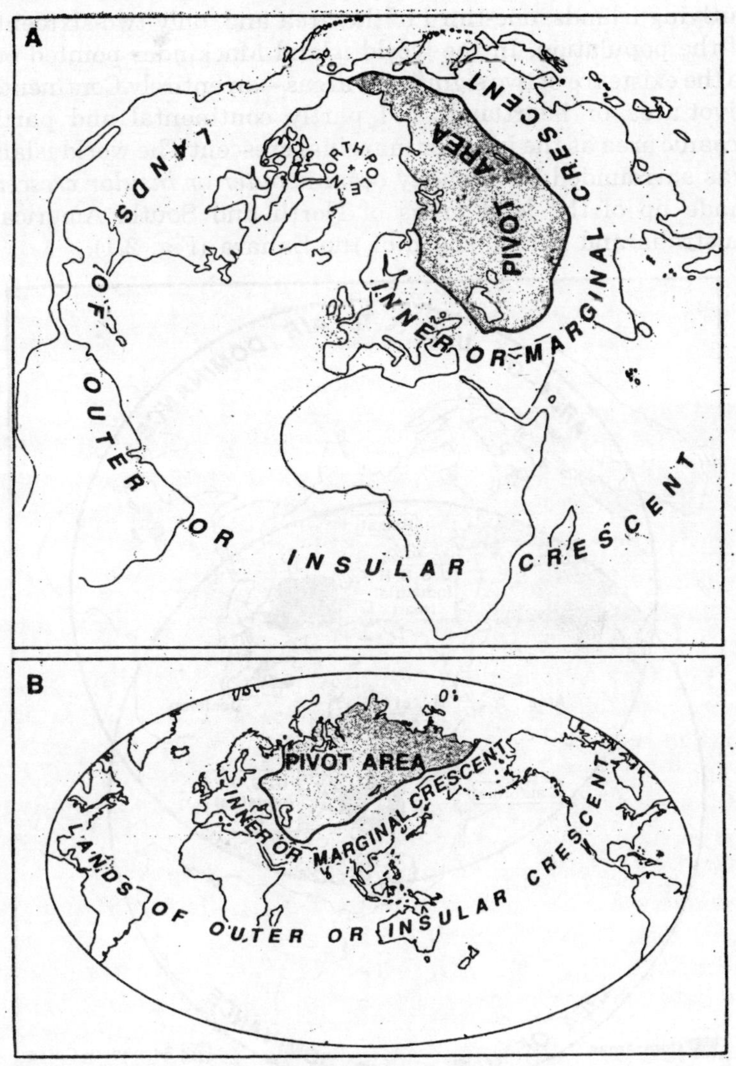

Fig. 2.1 : Mackinder's geopolitical organisation of the world: (A) as seen on a visual map projection; (B) as originally perceived on Mercator's map projection.

He considered the British and Japanese islands as *offshore islands* and the two Americas and Australia, as outlying islands. He noted that the world island covered two-thirds of the area and fourteen-sixteenth of the population of the world whereas outlying islands, one-third of the area and, only two-sixteenth of the population. In the world island Mackinder pointed out to the existence of two significant areas—an entirely Continental pivot-area or heartland, and partly continental and partly oceanic area as the inner or marginal crescent The world island was surrounded by a wholly oceanic *outer or insular crescent* made up of the land areas of North and South: Americas, Australia and Africa, south of the Sahara (Fig. 2.1).

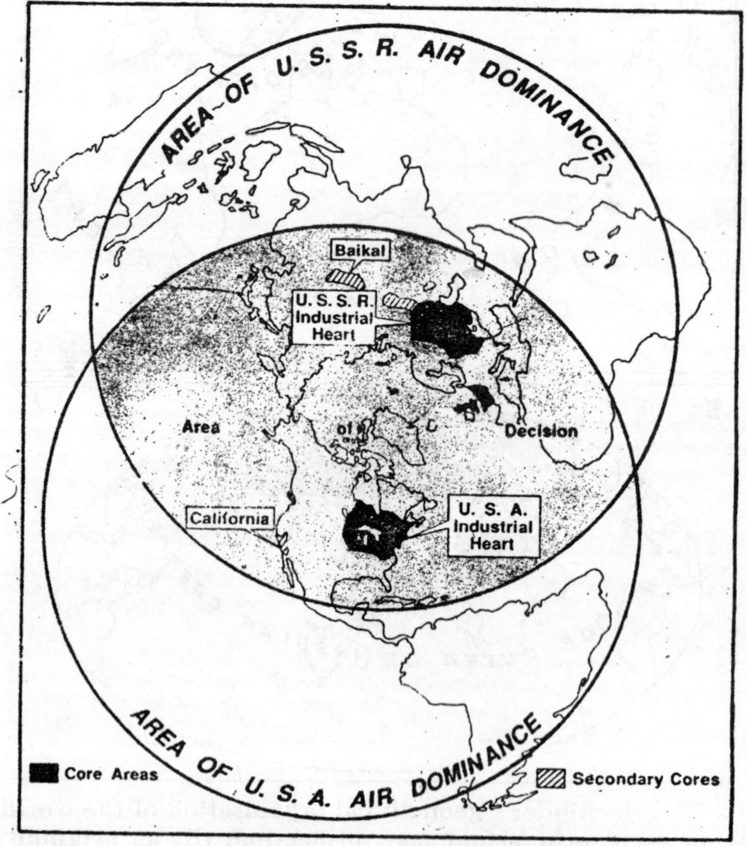

Fig. 2.2 : Airman's view of the world (de Seversky).

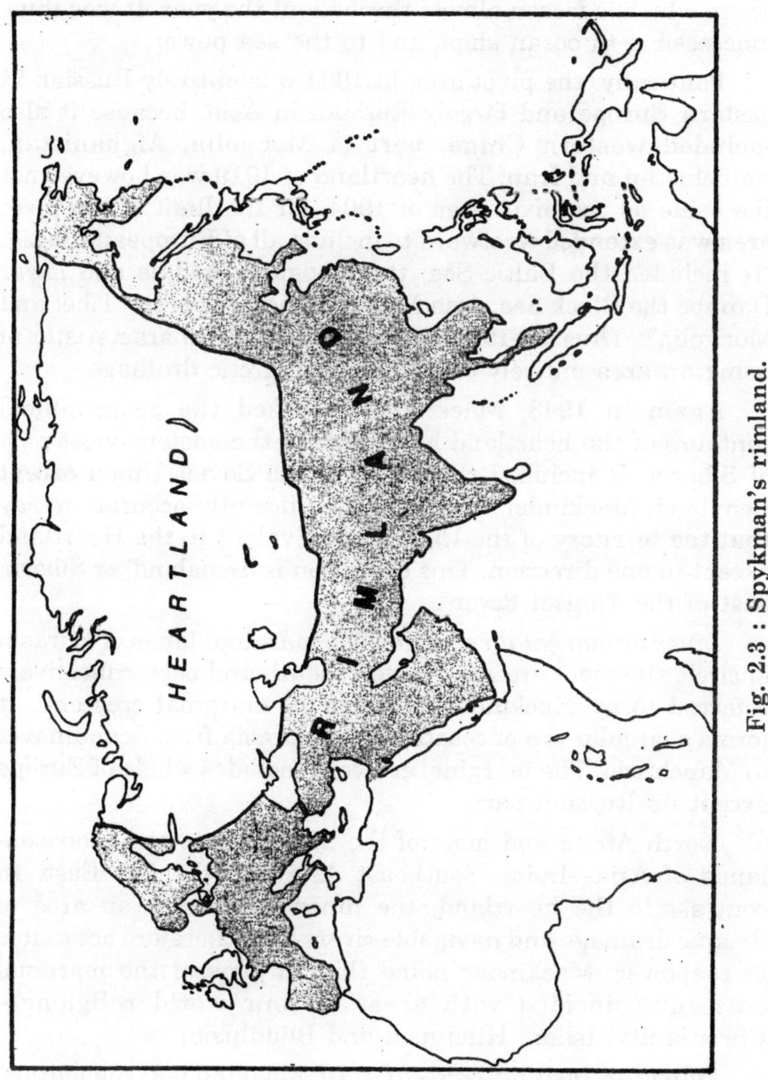

Fig. 2.3 : Spykman's rimland.

The Heartland : Mackinder defined the Heartland as a vast area in Eurasia, characterized by the Arctic and inland drainage. It stretched from the Volga, river to eastern Siberia and from the Himalayas to the Arctic Ocean and covered about nine million square miles or more than twice the area of Europe. It largely drained into inland seas and lakes and into the Arctic ocean which is frozen almost throughout the year. It was thus, inaccessible to ocean ships and to the sea power.

Politically, the pivot area in 1904 was entirely Russian in eastern Europe and largely Russian in Asia, because it also included western China, part of Mongolia, Afghanistan, Baluchistan and Iran. The heartland of 1919 was however not the same as the pivot area of 1904, for the limit of the pivot area was extended westward to include all of European Russia. "It included the Baltic Sea, the navigable middle and lower Danube the Black Sea, Asia Minor. Armenia, Persia, Tibet and Mongolia". Thus in 1919 heartland lost its characteristic of being an area entirely of "inland and Arctic drainage.

Again in 1943, Mackinder modified the geographical contours of the heartland by excluding the eastern-most part of Siberia. It included the whole of the Soviet Union except Lenaland. Mackinder wrote, "It is sufficiently accurate to say that the territory of the USSR is equivalent to the Heartland except in one direction. This exception is 'Lenaland' or Siberia east of the Yenisei River".

Inner or marginal crescent : The marginal lands of Eurasia encircle the pivot area on its west, south and east, collectively referred to by Mackinder as 'inner or marginal crescent'. It forms a circular are of coastlands of Eurasia from Scandinavia to Manchuria. The marginal crescent includes whole of Europe except its Russian part.

North Africa and most of the Middle East and monsoon-lands of Asia—India, Southeast Asia and the Far East. In contrast to the heartland, the inner crescent is an area of oceanic drainage and navigable rivers. It is therefore accessible to seapower. Mackinder noted that in general the marginal crescent coincided with areas of four world religions—Christianity, Islam, Hinduism and Buddhism.

Outer or insular crescent : In Mackinder's geopolitical organization of the world, beyond the inner crescent are located the world ocean and widely separated lands of the outer or insular crescent. It consisted largely of North and South Americas, and Australia, the British Isles, Japanese Islands, and Africa, south of the Sahara, which Mackinder called as the second heartland'.

Importance of Heartland

Mackinder attached greater importance to the heartland than either to the inner or the outer crescent. According to him heartland was a vast area of Arctic and interior drainage, surrounded by physical barriers on all sides except on the west which provided it the essential security from outside attack. The value of the heartland, in Mackinder's view, lay in its vast size, central location and, inaccessibility.

The vast size was the source of power and much needed self-sufficiency in food and raw materials, while centrality and inaccessibility increased the changes of defence. Mackinder hoped that the heartland would soon be covered with it railroad network which would replace the horse and camel as means of transport. He wrote, "The full development of railways was merely a matter of time. The spaces within the Russian empire and Mongolia are so vast and their potentiality in population, wheat, cotton, fuel and metals so incalculably great that a vast economic world more or less apart and inaccessible to oceanic commerce would inevitably develop' (Mackinder, 1904).

The Predictions

Mackinder believed that a great land power would emerge in the heartland and expand towards the marginal lands of Eurasia and from there to the ruler of the world. In 1904 he did not anticipate that the fear of world domination world come from the pivot state, Russia, but visualized instead that Germany might attempt to control the heartland and become "the ruler of world island and eventually of the world.

He pointed out that it was through the north European plain that the heartland could be penetrated by any outside power. He wrote in 1904, "the oversetting of the balance of power in favour of the pivot state, resulting in its expansion over the marginal lands of Euro-Asia would permit the use of vast continental resources for fleet-building and the empire of the world would then be in sight. This might happen if Germany were to ally herself with Russia".

But by 1943 he found it necessary to revise his original idea. In the Second World War both the U.S.A. and the U.S.S.R. demonstrated their increasing military strength and political significance. This made Mackinder to realize that the threat of world domination might come from the pivot state itself, namely, the Soviet Union. Before that Germany had made unsuccessful attempts to invade and conquer Russia. Mackinder therefore wrote, "All things considered, the conclusion is unavoidable that if the Soviet Union emerges from this war as conqueror of Germany, she must rank as the greatest land power on the globe. Moreover, she will be the power in the strategically strongest defensive position. The Heartland is the great natural fortress on the earth. For the first time in history it is manned by a garrison sufficient both in number and quality." He concluded that his heartland thesis was "more valid and useful today that it was either twenty or forty years ago". He re-examined the heartland concept in early forties and found the importance of this 'citadel of landpower' greater than ever.

While Great Britain and her allies paid little or no attention to the heartland theory between the two world wars, Germany found in it an argument for its policy of expansion towards the east and its attack on Russia in the early years of the World War H. It may be noted that on its publication in 1919 the heartland theory was ignored by all except a handful of people in the universities. But on its republication in 1944, it was read carefully by many people, specially in Britain and America, who offered a critical evaluation of the theory.

Critique of the Heartland Theory

It seems quite strange as to why the contemporaries of Mackinder did not question the logic of heartland concept at the time of its first statement in 1904. But in one way his theory immediately became the most intensely debated idea, since, the author of the theory himself modified the concept in 1919 and again in 1943. Several writers have criticized Mackinder for his inaccurate thinking as welt as theory. Prominent among them are Nicholas J. Spykman, Donald W. Meinig, David J.M. Hooson, Arthur R. Hall, Norman J. G. Pounds and W. G. Eaşt. While it is admitted that the theory has formed an important part of the basic reading in political geography, it has also aroused a lot of controversy and discussion. Mackinder has been praised and at the same time criticized.

"The heartland idea has inevitably drawn though rather late, shower of well-aimed brickbats as well as bouquets", specially after the Second World War.

Mackinder's predictions contained in his theory did not come true except that the former Soviet Union, the pivot State, exercised considerable influence over eastern Europe. His critics have assailed him for his failure to take into consideration the role of changing-technology. He has been criticized for over-emphasizing the potentialities and defensibility of the heartland.

It is true that navigation m the Arctic Ocean and the rivers that now into it is difficult and consequently the penetration of the heartland from the north was impossible for the seapower. Mackinder's world map on the Mercator projection immensely exaggerated the area of the frozen arctic Ocean and created a wrong impression that to the north of the heartland lies a large expanse of ice and ice only. But today the Arctic coast of Siberia presents less difficulty before the shipman, since the use of atomic ice breakers makes it and the Siberian rivers less forbidding to maritime contact.

Mackinder analysed the heartland concept in terms of land and sea power. Though he presented his 'Geographical Pivot of History' in 1904 at the beginning of the air age, yet he failed to recognize the significance of air power. Even in 1943 he was not convinced that air power would liquidate both fleets and armies. "But the standards of accessibility, mobility and strategic distance, crucial considerations in Mackinder's system, have become transformed: with, the tremendous growth of airpower and the arrival of inter-continental missile, and hydrogen bomb" (Hooson). Airpower can obviously achieve mass penetration into the heartland which sea power is or atleast was unable to do.

In 1950 Alexander de Seversky expressed the view that land and sea powers of the world have been subordinated to toe-greater power, of the air. His map on an azimuthal equidistant projection centred on the North Pole, shows very effectively the close proximity of the Eurasian Heartland from the one in Anglo-America across, the Arctic Mediterranean. It shows the areas of air dominance of the United States and the Soviet Union. The area of overtopping (Anglo-America, Eurasian heartland, maritime Europe, North Africa and the Middle East) is the area of decision, according to de Seversky (Fig. 2.2). The entire territory of one super power falls within the air dominance of the other.

This points to another major weakness of the heartland concept. De Seversky concluded that the great deterrent to heartland aggression should be the threat of large scale air retaliation by its rival power—the USA. Airpower has possibly upset the assumptions on which Mackinder's thesis was based. According to John Slessor the heartland concept has ceased to be relevant with the advent of the airborne nuclear weapons. "Air power, whose role is to supplement land and sea based forces, has turned the vast spaces that were her prime defence against Napolean, Hindenburg and Hitler into a source of weakness. The virtues of the Heartland size, centrality and inaccessibility have become either of no advantage or disadvantage" (Jones, 1955).

The defeat of Germany and Japan in the World War II and decline of Britain and France created a situation in which the former Soviet Union emerged as a dominant power. Yet A.R. Hall opined that the emergence of Russia did not substantiate Mackinder's prophecy because in his analysis he attached no importance to the presence of another heartland in Anglo-America, which is capable of playing as significant a role in world affairs as the one in Eurasia.

Some wrong assumptions have reduced the validity of the theory. For instance, "Mackinder confidently expected that the heartland would be covered with a network of rail-roads. This rail-road network has not yet been built even after the lapse of half a century" (Pounds). W.G. East has rightly said,. "In a world of rapid technological progress the geography of a territory, since it is ever changing, should be continually revalued. It is surely both unwise and dangerous to accept as a predetermined end the predictions that world hegemony must on certain assumption, inevitably pass on to the rulers of one specified portion of the earth."

Present-day Validity of Heartland Theory

It is clear from toe foregoing that Mackinder had based his predictions of world hegemony by either Germany or the former

Soviet Union on wrong presumptions. Today many people ask whether the" heartland theory of Mackinder is valid or not; whether the predictions which Mackinder made through his theory are somewhere nearer the truth or not. After the defeat of Germany in the second world war, the former Soviet Union emerged as a great power in the world. She succeeded in controlling the whole of Eastern Europe by installing communist governments in East Germany, Czechoslovakia, Hungary, Rumania and Poland. Since the former Soviet Union was the heartland state and East Europe went red, the first line of the theory, 'Who rules East Europe commands the Heartland', came to be true. The former Soviet Union ruled over East Europe by proxy and commanded the Heartland physically.

But the rise of the U.S.A. as a great power after the Second World War had also challenged the supremacy of the former Soviet Union. Everywhere in the world the former Soviet Union faced a stiff opposition from the U.S.A. Now the political situation of Europe has entirely changed. The ideological change of East Europe which is manifest is the overthrowing of communist regimes one after the other, the unification of both Germanics, brightening the prospects of the rise of a super state in Europe and the declaration of several republics of the Soviet Union as sovereign states, and other internal problems (economic etc.) considerably weakened the position of the former Soviet Union. Moreover, the challenging position of the U.S.A. reduced the importance as well as chances of the former Soviet Union for world domination.

Although, Russia still controls the heartland, she has lost her political grip over East Europe. Thus, the validity of even the first line of the theory is no more there. In the end it may be said that the heartland concept of Mackinder is not valid in the altered world situation of the early nineties.

The geographic relationships of politics have been a subject of continuing interest among scholars at least from the time of the early Greeks. Before the nineteenth century, however, the geographical perspectives were essentially local, since knowledge about the earth's surface was incomplete, and the means of transport and communication rudimentary. Owing to this, distances between places seemed intractable, so that it was incomprehensible to think of the world as a unit. For this reason, ideas about the geographic aspects of strategy were, confined mainly to the considerations of the impact of physical environment on politico-military manoeuvres. With the arrival of the Columbian era, sea-faring nations were beginning to establish distant colonial footholds, transport and communication were becoming faster, and the world was rapidly shrinking into a single family of diverse nations. The geographer's perspectives were changing accordingly. By the time of Ritter and Humboldt, geographers had begun to think of the entire surface of the earth as the unit of their study. This changed geographical perspective had begun to reflect in strategic writings by the end of the nineteenth century, particularly in the writings of Mahan. Later, it was put forward in a big way by Halford Mackinder in an epoch-making presentation before the Royal Geographical Society of London in 1904.

By the time of Mahan and Mackinder, the surface of the earth had been (almost) completely mapped and it seemed possible to discern certain seemingly persistent relationships between the political history of the world and the spatial pattern of global distribution of continents and water bodies. While scholars of the earlier generations had been attempting to explain the pattern of political structure in the world in terms of local and regional variations in environmental conditions, the new thinkers, beginning with Mahan and Mackinder, attempted to describe and explain world political patterns in terms of spatial models of world power, based on some principal (fundamental) causes inherent in history and geography.

They differed from their predecessors in that they tried to explain the contemporary world scheme by taking into consideration diverse factors, such as historical continuity, spatially shifting balance of world power, and technological revolution in communications and means of warfare. The common, and to some extent inevitable, shortcomings of the diverse "global strategic views" advanced since Mackinder put forward his heartland thesis in 1904, is that each thinker regarded the world political pattern prevailing in his time as final. The result was that as soon as a shift in balance of power occurred, or revolution in communication technology brought about changes in the existing pattern of space relationships, the proposed strategic models was rendered obsolete.

Communication technology has by far been the most potent factor in the developing views on global strategy. Until 1930s, the dominant technologies of communication were the oceanworthy ships, railway engines, and automobiles, so that places on the earth's surface could be reached only by surface transport over land or sea. Ideas about global strategy before 1940s, therefore, were speculated in terms of land power versus sea power dominance. During this phase, the world views of scholars and statesmen were shaped by a flat-earth perspective as defined by the popular Mercator map. Arrival of the aircraft changed the scenario completely. Air transport soon developed into the most critical factor in global contest for peace and war. The old concepts about global strategic relationships were revised and/or replaced by new ones, more hi keeping with the emerging reality. As the flat-earth world view was replaced by a round-world view, Mercator was dethroned, and the equidistant azimuthal projection centred on the North pole, began to define the parameters of thought on global relationships. In view of this, it appears logical that the following discussion be divided into flat-earth and round-world perspectives.

Another important point is that global strategic ideas should by no means be treated as objective realities, standing apart from their authors' location in history and geography. Every idea is a product of its time, and the space frame in which it originates. Howsoever objective the presentation may appear, the fact that the author is the citizen of a particular country, and has seen the world from that vantage point, imperceptively enters into his version of global politics. To take the most conspicuous example: Mackinder had presented the pivot-area/heartland thesis (1904) at a time when Britain was almost on the point of losing her predominant position in global politics. Her economic relations were dithering in the face of rising economic, industrial, and political power of .Germany, and the United States, in the closing years of the nineteenth century. Britain could no longer champion a liberal international economy based on unrestricted market and free trade. Ironically, Great

Britain herself needed protection for her industries from free entry of industrial products from Germany. From this perspective, the Mackinder theory may be interpreted as the historical and geographical rationalization for the perpetuation of the traditional British policy of balance of power in Europe, in order to ensure that no single continental power (or a combination of them) could pose a threat to Britain's leadership. A major concern was to prevent any alliance between Germany and Russia so that Germany may not get access to the pivot-area and thereby, to the vast resources of a continental empire. The inherent fear was that Germany could supersede Britain by virtue of her richer resource base. British foreign policy needed a new initiative, a new outlook, more in keeping with the changed politico-economic scenario. Mackinder offered one such outlook (Tuathail, 1992). Such a perspective should explain why working out from similar historical-geographical premises, Haushofer and Spykman had reached entirely different conclusions and policy formulations for their respective countries.

GLOBAL STRATEGY IN THE AGE OF SURFACE TRANSPORT : THE FLAT-EARTH PERSPECTIVES

Geostrategic Ideas of Mahan

No individual has so directly and profoundly influenced the theory of sea power and naval strategy as Alfred Thayer Mahan (1840-1914). "He precipitated and guided a long-pending revolution in American naval policy... In one way or another his writings affected the character of naval thought in France, Italy, Russia, Japan and lesser powers... By direct influence and through the political power of his friends—Theodore Roosevelt and Henry Cabot Lodge—he played a leading role in persuading the United States to pursue a larger destiny overseas during the opening years of the twentieth century" (Margaret Sprout, 1943, p. 415).

Born in 1840, Mahan graduated from the United States Naval Academy in 1859. He spent 41 years of active service with the US Navy and retired from it in 1906, with the rank of Rear Admiral: At the age of 44 he joined the Naval War College at Newport, of which he became the President in 1886.

The American Historical Association elected him as its president for the year 1902-1903 in recognition of his contributions to naval history.

Mahan's ideas about global strategy cannot be found in any single book or article. These have to be pieced together from the numerous statements scattered throughout his large number of writings. His major contribution consists of the great trilogy of the following books. *The Influence of Sea Power upon History, 1660-1783* (1890); *The Influence of Sea Power upon the French Revolution and Empire*, 1793-1832(1892); and The Life of Nelson (1897). All his writings are linked through a common cord derived from his firm conviction that the essential condition for world power status for any State was an effective control of the seas. He was convinced that in the struggle for supremacy among the leading States of the world, the oceanic powers would always have the upper hand.

The following four spatio-political aspects on the role of sea power in world politics were basic to the strategic thinking of Mahan:

1. There is a continuous and unbroken system of oceans and seas on the earth's surface leading to a unified system of oceanic communication.
2. There is a vast trans-continental and almost completely landlocked State—the Russian Empire extending without a break from the ice-bound Arctic in the north to the rugged desert-mountain belt of inner Asia in the south, and from eastern Europe to the extreme eastern margin of the Asian mainland.
3. Around the trans-continental Russian Empire are the maritime States of Europe and the countries of South and East Asia, bordering on the Indian and the Pacific oceans.
4. There are three insular States disconnected from the Eurasian mainland—Great Britain, Japan and the United States.

To Mahan, the sea was a great highway; "a wide common, over which men may pass in all directions, but on which some well-known paths show that controlling reasons have led them to choose certain lines of travel rather than others". In a trade dominated modern world, therefore, suitable oceanic location offered a distinct politico-economic advantage, and a landlocked location was in that regard, a definite disadvantage. Mahan wrote that,

> the ships that... sail to and fro must have secure ports to which to return, and must, as far as possible, be followed by the protection of their country throughout the voyage... In these three things—production, with the necessity of exchanging products; shipping, whereby the exchange is carried on; and colonies, which facilitate and enlarge the operations of shipping and tend to protect it by multiplying points of safety—is to be found the key to much of the history, as well as of policy, of nations bordering the sea.

To Mahan, the Russian Empire typified the strength and the weakness of land power. Obstacles imposed by physical geography placed a strong check upon the freedom of movement of the Russian State. Enormous distances and inadequate internal communication severely limited its internal coherence. Because of its landlocked position, Russia was also greatly disadvantaged in that it could be completely shut off by a hostile sea power. Owing to these difficulties inherent in her geography, Mahan thought that the Russian Empire, if it could not be broken it could at least be contained.

In sharp contrast, was the geographical position of Great Britain. Despite her small home base, she had established a worldwide empire, by virtue of her advantageous sea frontage, that helped her to emerge as the strongest sea power and, hence also the leading world power. Irrespective of her transcontinental spread and huge size, the Russian Empire could easily be contained and immobilized by the strong British navy with its far-flung bases.

Mahan closely studied and analyzed the basic features of Britain's world-wide sea power. He noted that among the European powers, Britain alone was an insular country. By virtue of her insularity, Britain was, in a large measure, secure from outside attack. It needed no heavy outlay on defence. This made it possible for her to save manpower and money for investment in the navy. Britain's geographical location between the northern European country and the Atlantic, conferred on her immense strategic advantages in trade and commerce, since deep-sea commerce of all European ports lying to the north of London had to pass through the English Channel. British naval forces could, therefore, block the deep-sea commerce of the whole of northern and western Europe. Taking advantage of her oceanic location, Britain succeeded in establishing control over most strategic points on nearly all the leading routes of world trade. Control over the Strait of Gibraltar, the Suez Canal, the Cape of Good Hope, Singapore, Hong Kong, the Strait of Magellan, and Cape Horn gave Great Britain, a commanding position in Europe, Asia, and the Pacific. From Nova Scotia, the British navy flanked the great circle route from the United States to the Far East, while the British West Indian islands, offered a cover over the trade passing through the Panama Canal.

According to Mahan, among the States of the world, only the United States had a geographical position comparing or even rivalling that of Great Britain. It had no powerful enemies on its borders; like Britain it was also insular and hence possessed all the advantages of defence and security that insularity offers. Although it lacked valuable colonies, it was more than adequately compensated by its easy and secure access to the vast resources of nearly half the North American continent. Besides, her position between the two great oceans—the Atlantic and the Pacific—offered her further ad- vantages as a sea power. Mahan was a great advocate of the opening of the Panama Canal across the Panama Isthmus, connecting the Pacific and the Atlantic sea-boards of the United States. In his later years, Mahan was fully convinced that this great nation (*i.e.*, USA) would replace Britain as the leading sea power.

Determinants of Sea Power

From his analysis of the British sea power, and the naval history of the world, Mahan was able to identify a set of six factors which, in his opinion, largely determined sea power:

1. *Geographical Location of the State* : This included location in respect of (i) frontage on one or more seas, and (ii) the extent of land frontiers.

2. *Physical Layout of the State :* According to Mahan: The seaboard of a country is one of its frontiers; and easier the access offered by the frontier to the region beyond, in this case the sea, the greater will be the tendency of a people toward intercourse with the rest of the world by it. If a country be imagined having a long seaboard, but entirely without a harbour, such a country can have no sea trade of its own, no shipping, no navy.... Numerous and deep harbours are a source of strength and wealth, and doubly so if they are the outlets of navigable streams, which facilitate the concentration in them of the country's internal trade; but by their very accessibility they become a source of weakness in war, if not properly defended (Mahan, 1890 [1949 ed.], p. 25).
3. *Extent of Territory :* This included the length of the country's coastline and its defensibility against foreign invasion.
4. *Size of Population* Mahan was convinced that only a populous State could build and man a strong navy. Only such a State could maintain overseas bases and colonies, which required a sizeable force of soldiers and administrators.
5. *National Character* By this, was implied, a favourable aptitude for seafaring and commerce.
6. *The Character and Policy of Government* According to Mahan, a farsighted government with a bold administrative policy is required in order that the best possible use can be made of the advantages offered by natural location and large and an enterprising population.. Mahan paid special attention to the politico-geographical situation in Asia.

In his view, the Russian strategy was to drive a wedge between Afghanistan and Iran to reach the shores of the Indian Ocean, while in the Far East, it was trying to push southwards through Manchuria to the narrow seas opening into the Pacific. Against this advance, the only effective check that maritime powers could place was to provide a strong military defence along the periphery of the Russian Empire, since its geographical location seemed to have made it a kind of invulnerable fortress. "The Russian centre", he wrote, "cannot be broken". It was imperative, therefore, that the routes of advance from the "centre" should be blocked. According to Mahan, the struggle against the Russian colossus should be confined mainly to the southern and eastern borderlands of Asia, since,

> a grand strategy of blocking Russian expansion was in the interest not only of the insular nations but also of the maritime nations of continental Europe. That such an oceanic coalition could achieve its objective of containing Russian expansion; he had not the slightest doubt. The combined resources of Britain, Japan, Germany and the United States, he argued, 'outweigh—in foundation and in superstructure' the potentialities of Russia, even in alliance with France (Sprout and Sprout, 1964, pp. 323-324).

In the opinion of Harold and Margaret Sprout, in his geopolitical analysis of the contemporary' world, Mahan

> clearly anticipated Mackinder's concept of the Eurasian 'Heartland'. His strategy vis-a-vis the Russian Empire closely resembled in geopolitical conception the strategy for containing the Soviet Union [that was] actually pursued by the United States during the later 1940s and 1950s, nearly half a century after Mahan's death in 1914 (Sprout and Sprout, op. cit., p. 325).

However, like all other strategic speculations, Mahan's also was written in the perspective of its time; a time when the navy of Queen Victoria (1837-1901) reigned supreme. Now that the technology of transportation and warfare have undergone revolutionary changes, and air power has in a good measure replaced sea power, much of what Mahan wrote has become obsolete.

But this does not lessen the significance of his writings, which were full of great insight, and contributed a great deal to mold the strategic thinking of his countrymen. And with the arrival of a new oceanic age, heralded by the nuclear powered sub-surface ship, Mahan might become relevant once again.

Geostrategic Ideas of Halford Mackinder

Sir Halford Mackinder (1861-1947) has perhaps been the best known geographer in the non-geographic circles, and his has definitely .been the best known global strategic scheme in modem times. But his views have not necessarily been the most respected. Many believed that, "It would be difficult to find an essay of comparable length and reputation that is more indefensible in terminology and ideas than Mackinder's" (Malin, 1959, pp. 339-363). But this is generally true of most other ideas that area rendered out of date by the rapid march of events and technology.

A Scot by birth, Mackinder was professor of geography at the University of London, Director of the London School of Economics and a Member of the British Parliament. In a more organized and pointed way than Mahan, Mackinder provided a fresh political perspective on the geographical distribution of land masses and water bodies on the earth's surface. He interpreted history as essentially a struggle between land and sea power. The geopolitical concepts and hypotheses of Mahan and Mackinder are by no means as far apart as has often been asserted. Drawing a contrast between their theories of sea and land power is a false antithesis, since,

> their basic propositions are in certain respects so similar as to revoke at once a conjecture that Mahan and Mackinder must have avidly read each other's books and magazine articles—But it was Mackinder rather than Mahan who methodologically ordered the concepts and hypotheses, propounded an interpretation of history in the grand style, and drew a geopolitical blueprint of the future which for better or for worse still grips the imagination of men in many lands (Sprout and Sprout, 1964, p. 320).

Fundamentals of Mackinder's Concept of Pivot-Area (1904)

In 1904, Mackinder presented before the Royal Geographical Society of London, his seminal paper entitled "The Geographical Pivot of History". It is in this paper that the basic ideas of his "heartland" thesis were, for the first time, put forward in an organized manner. Mackinder wrote that in the opening decade of the twentieth century the geographers were for the first time in a position to attempt, with some degree of completeness, a correlation between the larger geographical and the larger historical generalizations, and could seek therefrom, a formula of "geographical causation in universal history". This seemed possible to him because in the post-Columbian era, international politics had become a closed system of worldwide scope. Mackinder presented his paper in order to elucidate his formula of "geographical causation" in world history.

From his study of the physical map of the world, Mackinder concluded that on the global scale, the earth surface consisted of a huge and continuous landmass, consisting of Europe, Asia, and Africa, and some isolated islands, *e.g.*, North and South America, Australia, Great Britain, and Japan. He named this huge united landmass of Europe, Asia and Africa as the "World Island". Mackinder noted that three-fourths of the earth surface is water and only one-fourth land. Of this one-fourth, two-thirds are accounted for by the World Island and only a third is covered by the remaining continents. The World Island also accounted for seven-eighth of the total population of the world. The rest of the continents supported only one-eighth of mankind. He envisioned the landmasses of the world arranged in three tiers (Fig. 2.4).

In the first tier, there is a huge area of interior and Arctic drainage located in inner Eurasia. This is surrounded by mountains on three sides and by the ice-bound Arctic on the fourth. This was his so-called "pivot area" that he later named as the "Heartland". It stretched from the Volga river in the west to eastern Siberia in the east, and from the Himalayas and the related mountains in the south to the Arctic in the north. The distinguishing feature of Mackinder's Heartland area was that since it was surrounded by mountain barriers on three sides and an ice-bound sea on the fourth, it was not accessible to sea powers and was, therefore, strategically secure like a fortress. The Heartland was, however, vulnerable to land forces in the southwest, through eastern Europe, between the Ural mountains and the Caspian Sea, where a narrow corridor, covered by steppe grasslands, offered an easy entry (Fig. 2.5).

The "pivot area" was pictured by Mackinder as surrounded by an "inner" or "marginal" crescent, consisting of an arch of coastland, and characterized by drainage into navigable seas. This Inner or Marginal Crescent consisted of the whole of Eurasia outside the Heartland. It included Europe west of the Ural mountains, southwest Asia outside the Iranian upland, India, South-East Asia, and most of China.

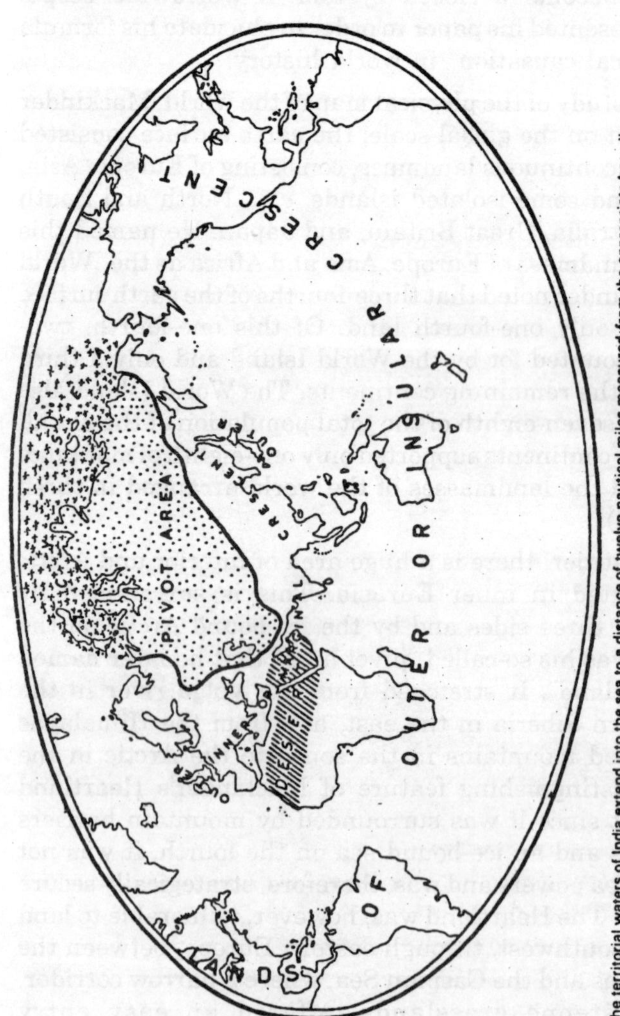

The territorial waters of India extend into the sea to a distance of twelve nautical miles measured from the appropriate base line

The territorial waters of India extend into the sea to a distance of twelve nautical miles measured from the appropriate base line

Fig. 2.4 : Mackinder's Heartland model of 1904—The Natural Seats of Power: Pivotal-area—wholly continental; outer crescent—wholly oceanic; Inner crescent—partly oceanic, partly continental.

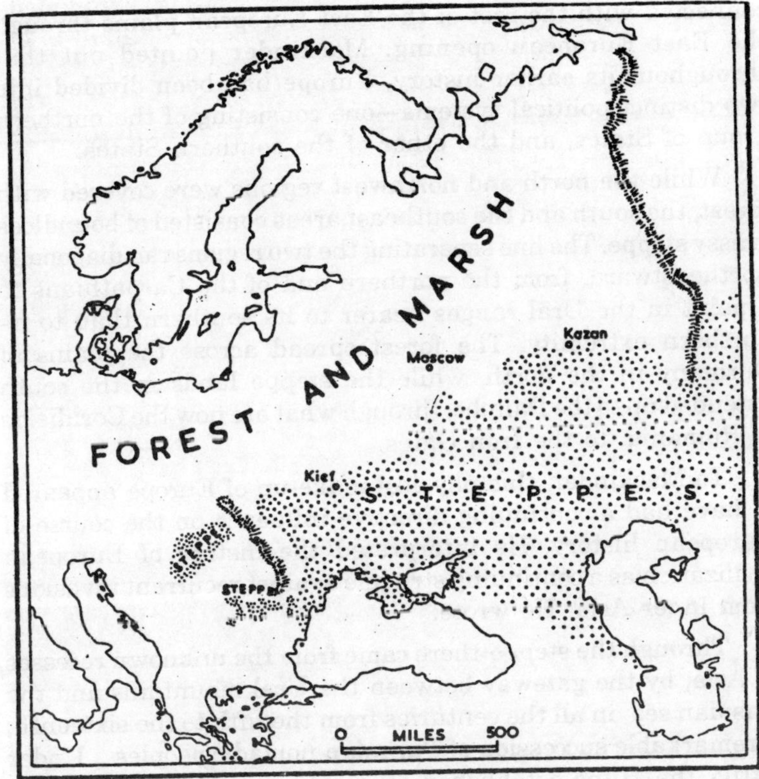

Fig. 2.5 : Mackinder's "Pivot Area" The East European Corridor (after Mackinder, 1904).

Outside the Inner or Marginal Crescent lay his third tier which he named as the Outer or Insular Crescent. Besides the continents of North and South America, Australia, and Africa south of the Sahara, the island regions of Great Britain and Japan were also designated as "outer" because of their complete insularity from the mainland of Eurasia.

From his study of Eurasian history as well as the contemporary political map of the world, Mackinder was convinced about the great strategic importance possessed by the inner Eurasian region in the affairs of the Old World. The north, central, and western parts of the Heartland constituted a vast lowland, broken only by the Ural Mountains, and were connected with the rest of the East European plains through the East European opening. Mackinder pointed out that throughout its earlier history, Europe had been divided into two distinct political systems—one consisting of the northern group of States, and the other of the southern States.

While the north and northwest regions were covered with forest, the south and the southeast areas consisted of boundless grassy steppe. The line separating the two regions ran diagonally northeastward, from the northern end of the Carpathians to a point in the Ural ranges nearer to its southern than to its northern extremity. The forest spread across the plains of Germany in the north, while the steppe lands in the south extended up to the Danube, through what are now the Cornfields of Rumania, to the Iron Gates.

To Mackinder, this physical make-up of Europe appeared to have had a continuing directive influence on the course of European history. He interpreted the history of European civilization as a continuing struggle against recurrent invasions from inner Asia. He wrote:

Through the steppe-there came from the unknown recesses of Asia, by the gateway between the Ural mountains and the Caspian sea, in all the centuries from the fifth to the sixteenth, a remarkable succession of Turanian nomadic peoples... Under Attila the Huns established themselves in the midst of the Pusstas, the outer most Danubian outlier of the steppes, and thence dealt blows northward, westward, and south-ward against the settled peoples of Europe. A large part of modem history might be written as a commentary upon the changes directly or indirectly ensuing from these raids...

> The hordes which ultimately fell upon Europe in the middle of the fourteenth century gathered their first force 300 miles away on the high steppes of Mongolia... While the Golden Horde occupied the steppe of Kipchak, from the Sea of Aral, through the interval between the Ural range and the Caspian, to the foot of the Carpathians, another horde descending south-westward between the Caspian sea and the Hindu Kush into-Persia, Mesopotamia, and even into Syria, founded the domain of IIkhan. A third subsequently struck into

> northern China, conquering Cathay. India and Mangi, or Southern China, were for a time sheltered by the incomparable barrier of Tibet, to whose efficacy there is, perhaps, nothing similar in the world... But at a later time, in the days of Marco Polo in the case of Mangi, in those of Tamerlane in case of India, the obstacle was circumvented. Thus... all the settled margins of the Old World sooner or later felt the expansive force of mobile power originating in the steppe. Russia, Persia, India, and China were either made tributary, or received Mongol dynasties (Mackinder, 1904, pp. 425-430).

In the pre-Columbian age of overland transport and communication, this "heartland" region of steppe grasslands, except in the sub-Arctic forest, was very favourable to the mobility of horsemen and camelmen. As such, this region possessed a distinct strategic advantage over the lands of the Marginal Crescent. It was this that made possible the "cloud of ruthless and idealess horsemen sweeping over the unimpeded plain—a blow, as it were, from the great Asiatic hammer striking freely through the vacant space".

Mackinder noted that mobility upon the ocean was the natural rival of the horse and camel mobility in the heart of the continent. The discovery of the Cape route to India and the Far East, had helped to connect the western and eastern coastal navigations of Euro-Asia, and thereby neutralized the strategical advantage of the steppe-nomads. The broad political effect of the revolution commenced by the great mariners of the Columbian generations, was "to reverse the relations of Europe and Asia". As Mackinder put it, "whereas in the Middle Ages Europe was caged between an impassable desert to south, an unknown ocean to west, and icy or forested wastes to north and north-east, and in the east and south-east was constantly threatened by the superior mobility of the horsemen and camelmen, she now emerged upon the world, multiplying more than thirty-fold the sea surface and coastal plains to which she had access, and wrapping her influence round the Euro-Asiatic land power which had hitherto threatened her very existence" (Mackinder, 1904, p. 433).

Mackinder was, however, convinced that this shift of balance was only a temporary affair for, he thought that, "transcontinental railways are now transmuting the conditions of land-power, and nowhere can they have such effect as in the closed heartland of Euro-Asia, in vast areas of which, neither timber nor accessible stone was available for road-making. Railways work the greater wonders in the steppe, because they directly replace horse and camel mobility, the road stage of development having here been omitted". He thought that "the century will not be old" before the whole of Asia would be covered with railways. He wrote:

> The spaces within the Russian Empire and Mongolia are so vast, and their potentialities in population, wheat, cotton, fuel, and metals so incalculably great, that it is inevitable that a vast economic world, more or less apart, will there develop inaccessible to oceanic commerce... There have been and are here the conditions of a mobility of military and economic power of a far-reaching and yet limited character. Russia replaces the Mongol Empire. Her pressure on Finland, on Scandinavia, on Poland, on Turkey, on Persia, on India, and-on China replaces the centrifugal raids of the steppemen. In the world at large she occupies the central strategical position held by Germany in Europe. She can strike on all sides and be struck from all sides, save the north. The full development of her modem railway mobility is merely a matter of time (Mackinder, 1904, pp. 435-436).

Mackinder was emphatic that the likely "over-setting of the balance of power in favour of the pivot State, resulting in its expansion over the marginal lands of Euro-Asia, would permit the "use of vast continental resources for fleet building, and the empire of the world would then be in sight". Since at that time the Russian State, by itself, did not appear to Mackinder to be strong enough to be able to convert this possibility into reality, he envisaged two possible ways in which this could happen. First, if Germany were to ally herself with Russia; and secondly, if the Chinese, supported by the Japanese, were to overthrow the Russian Empire and conquer its territory. Concluding his

1904 essay, Mackinder wrote that in the event of the second possibility, the Chinese "might constitute the yellow peril to the world's freedom just because they would add an oceanic frontage to the resources of the great continent, an advantage as yet denied to the Russian tenant of the pivot region".

Mackinder was clear in his mind that a sea power could, in the long run, be no stronger than the land base that sustained it (*i.e.*, its home base). Although over a period of three centuries, Britain had succeeded in building a world-wide empire, and a global sea power, in Mackinder's opinion, in the twentieth century, Britain was the legatee of a depreciating estate for, he wrote, "in the presence of vast powers broad-based on resources of half-continents, Britain could not again become mistress of the seas" (Mackinder, 1902, p. 358).

The Modified Concept of "Strategic Heartland", 1919

In 1919, Mackinder published a somewhat laboured and extended version of his 1904 paper in a book form entitled, *The Democratic Ideals and Reality*. In this book he renamed the "pivot region" as the Heartland. He wrote this book especially with a view to warn and instruct the statesmen of the Allied countries. The Allied powers, at that time were engaged in discussing the terms of peace with Germany which had been defeated in the First World War. The facts agitating Mackinder's mind were that the Russian Empire had already collapsed in 1917, and her new Communist rulers had made peace with Germany almost on her own terms. Events of the First World War, like when the navy of Great Britain had failed to force entry through the Turkish straits into the Black Sea, and the German mines had excluded them from the Baltic Sea, made it clear to Mackinder that sea power was capable of controlling a much less extensive area of the World Island than he had originally believed. He was now convinced that from the viewpoint of strategical thinking, the Heartland was much larger than the one originally defined by him in terms of physical geography. He redefined the Heartland as "the region to which, under modern conditions, sea-power can be refused access, though the western part of it lies without the region of Arctic and Continental drainage". His modified "strategic heartland" included the Baltic Sea, the navigable middle and lower Danube, the Black Sea, Asia Minor, Armenia, Persia, Tibet, and Mongolia (Fig. 2.6).

Furthermore, in 1917, the German armies had advanced through the steppe corridor of south-western Russia. This seemed to Mackinder to re-emphasize the strategic importance of the East European gate of entry into the Heartland. Convinced that the East European corridor offered the only key to control the Heartland by an outside power, he discounted China and Japan, and emphasized that the real threat of the future control of the Heartland by an outside power lay only from the side of a strong and ambitious Germany. Addressing the Allied Powers he wrote: "Unless you would lay up trouble for the future, you cannot... leave such a condition of affairs in East Europe and the Heartland, as would offer scope for ambition in future, for you have escaped too narrowly from the recent danger." He wished that some "airy cherub" could keep whispering in the ears of the Allied statesmen at the Peace Conference, his strategic formula:

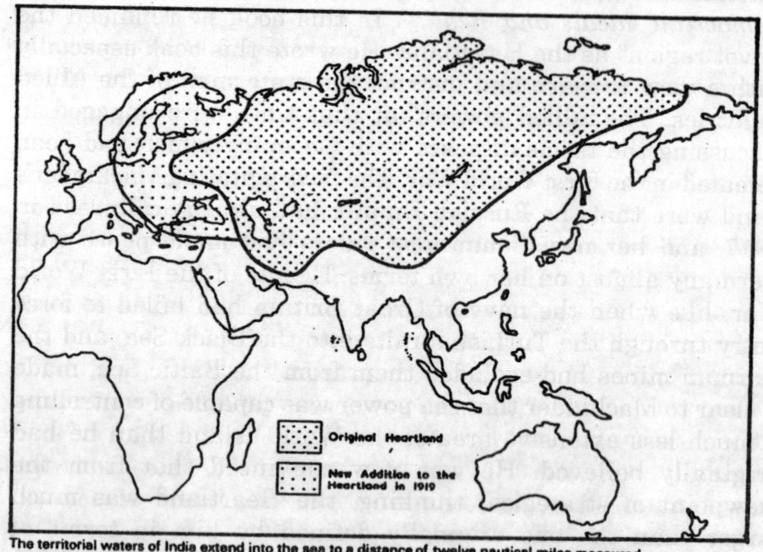

The territorial waters of India extend into the sea to a distance of twelve nautical miles measured from the appropriate base line

Fig. 2.6 : Mackinder's "strategic" Heartland of 1919.

Who rules East Europe commands the Heartland:

Who rules the Heartland commands the World-Island:

Who rules the World-Island commands the World (Mackinder, 1919, p. 113).

The strategic argument is simple. The Heartland is the strongest fortress on the earth, commanding resources of a huge trans-continental area. Any power that could organize it effectively was, in Mackinder's view, bound to emerge as a great colossus in world politics, head and shoulders above any other State. It, therefore, held the key to world supremacy. But this fortress was open in the southwest through the steppe corridor. This gate of entry could be controlled only by a power supreme in Eastern Europe, since Russia at that period, was not by itself strong enough to firmly control and manage the Heartland. Any outside power that could control and develop the Heartland into a powerful State, was bound to emerge far stronger than any other individual State in the whole of the Inner or Marginal Crescent. It could subjugate each of the States of the Marginal Crescent one by one and thus emerge as the ruler of the World Island. And any power that ruled the World Island would thereby command two-thirds of the land area and seven-eighths of the population of the world. As such, there could hardly be any doubt regarding the capacity of such a power to establish a world-wide empire.

The Modified Concept of 1943: The Midland Basin

Until 1919 Mackinder had confined his attention to Eurasia and the Old World. By the mid-1920s, it began to dawn upon him that Western Europe and North America constituted for all practical purposes, a single community of nations. He wrote: "In the United States, the most abundant rainfall and the most productive coalfields are to be found in the east, but in Europe they are in the west. Thus the west of the Europe and the east of North America are physical complements of one another and are rapidly becoming balanced halves of a single community" (Mackinder, 1924, p. 252). This changed emphasis in his perception of the politico-geographic reality was to mirror in yet another major modification of his global strategic view. This appeared as an article in the Foreign Affairs in 1943.

In this article entitled, "The Round World and the Winning of the Peace," and written at the time when Germany was on the verge of virtual annihilation, Mackinder no longer contended that "who rules the Heartland commands the World Island, and who rules the World Island, commands the World". He now clearly recognized (what he had failed to do in 1919) that the two sides of the North Atlantic were bound together by mutually supporting links of sea and air communications, and that the combined industrial and military strength of the United States and Western Europe was a force to be reckoned with. He described this geostrategic area comprising the North Atlantic Ocean, the eastern United States, and Western Europe, as the "midland basin".

He regarded this area as an effective counterbalance to the emerging political power-potential of the Eurasian Heartland. In 1943 he excluded from his Heartland, that part of the Soviet Union lying to the east of the Yenisei River. He named it as Lenaland. Owing to its rugged topography and forest cover, this region was of little economic value. Mackinder noted that the Second World War had drastically changed the politico-geographic picture of the world. He wrote: "If the Soviet Union emerges from this war as the conqueror of Germany [as she actually did], she must rank as the greatest land power on the globe. Moreover, she will be the power in the strategically strongest defensive position. The Heartland is the greatest natural fortress on earth. For the first time in history, it is manned by a garrison sufficient both in number and quality" (Mackinder, 1943, p. 595-605).

What he meant to say was that the War and the shattering defeat of Germany had turned the Heartland region from a power vacuum to a power centre.

Mackinder now visualized the emergence of two great centres of power in the world—the Midland Basin and the Heartland—separated by the German realm of East and Central Europe. He envisioned the two centres of power to be surrounded, and insulated, by a girdle of deserts—Saharan, Arabian, Iranian, Tibetan, and Mongolian—extending through the rugged and desolate Lenaland to Alaska, the Canadian Arctic, and the deserts of western America. Mackinder thought that this insulating tier of vacant land-spaces would for a "long time to come—break social continuity between major communities of mankind on the globe," and should thereby prevent the spread of conflict to the outside world. This was, however, not to be,' as the spread of great power rivalry to the farthest corners of the world was soon to prove.

Mackinder's Heartland: A Critical Appraisal

Although Mackinder has been the greatest influence on global strategic thinking in the modern times, he has nevertheless been variously criticized, both in respect of his facts and ideas. In the first place, he oversimplified history in a rather deterministic fashion as a struggle between land and sea power, which in fact was far from the truth. Secondly, Mackinder wrongly equated power potential with sheer geographical area, a factor that made him to over inflate the resource and power potential of the inner Asian Heartland. Thirdly, Mackinder overlooked the fact that the Heartland was a region of permanent difficulties because of its interior location and, owing to that, the extremes of climate. It is too cold in winter, and hot and dry in summer. Owing to these locational-climatic factors, the Heartland is at a relative disadvantage as compared to certain other larger and well-endowed areas of the world. Fourthly, he took a rather simplistic view of the possibilities of the development of railways in the sparsely settled Heartland region.

His expectations in this regards have not come true. Fifthly one of the greatest shortcomings of his thesis lay in that he tried to explain the present, and prognosticate the future, in terms of a technology that had already become outmoded. He wrote at a time when, in a general sense, the railway age was ending and the air age starting. Still he failed to take into account this new form of mobility which converted the sky into a highway, and thereby, cut into the very basis of his theory regarding the Heartland as an invincible fortress. The fortress now lay open to air raids from all sides, as much as any other place.

Another serious consequence of Mackinder's thinking about the present in terms of the past technology was that at the beginning of the twentieth century he was still explaining global strategy in terms of a Mercator projection map which created a false picture regarding the unlimited expanse of Arctic ice separating North America from Eurasia, The reality is that North America and the Heartland lie face to face across the narrow expanse of the Arctic. Since the shortest air routes connecting the Soviet Union and the United States lie across this narrow sea, the Arctic had got transformed as the Mediterranean of the modern times (Weigert, 1949). (In a joint paper in 1943, Harrison and Weigert showed that although it was depicted as enclosed within an ellipse, Mackinder's strategic map was actually drawn on the Mercator projection.) (See Fig. 2.11).

With the rapid march of science and technology, surface configuration of the earth, (*i.e.*, the global distribution of landmasses and water bodies), has become very much less significant politically than it was at the time when Mackinder first presented his "pivot area" thesis. Indeed, "When ballistic rockets armed with thermonuclear warheads can be fired from nearly any point upon the earth's surface, we... have reached the end of the line of geopolitical concepts and theories which purport to explain and to forecast the overall design of international politics by reference to the configuration of lands and seas" (Sprout and Sprout, 1964, p. 338).

VARIATIONS ON THE HEARTLAND THESIS

Mackinder's ideas on global strategy have been so much a part of modern strategic thinking in the non-Communist world that a number of writers have presented modified versions of his original thesis. Some of these may be briefly mentioned.

James Fairgrieve and the "Crush Zone" Concept

The first major variation on the heartland thesis was presented by British geographer, James Fairgrieve. In his Geography and World Power (1915), Fairgrieve gave his own interpretation on the distribution of land and water bodies on the earth surface, as viewed in the backdrop of the history of civilization in the Old World. According to this view, the centre of the world lay in the "old world parallelogram", which he visualized as being cut diagonally into two by a belt of low-lying and submerged land area. (Fig. 2.7). Fairgrieve underlined that this belt of lowlands was the main hub of the history of civilization. It was by the powers based in this belt that the coastal regions of the Old World paralleogram had been conquered and brought into the worldwide maritime system. He also drew attention to the existence of a strategic "heartland" in "the great plain of the world," located in central Eurasia, and the home of warlike pastoral nomads who had, all through history, made intermittent forays into the settled margins of the Old World.

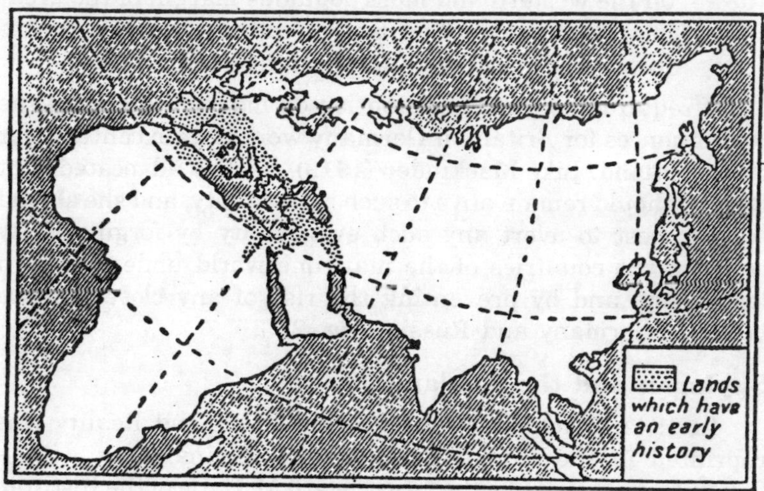

Fig. 2.7 : Fairgrieve's Old World parallelogram (Source: Fairgrieve, 1915)

Fairgrieve conceptualized a three-tiered arrangement of the combined land mass of Europe, Asia, and Africa. These were: the Heartland, the land of the sea powers, and the "crush zone," consisting of a narrow belt of land enclosed between the former two. The distinctive feature of the Crush Zone is that it is here that the Maritime and the Continental worlds meet.

Owing to this fact, all through history, this region remained under constant pressure from the rival groups on its two sides. As such, the Crush Zone was, in many ways, the most critical region in any contest for supremacy between the Maritime and the land based powers.

Fairgrieve's book had for a long time remained the standard text on this subject; it went through several editions until 1930s. In the post-First World War editions of his book, the author underlined the critical role that the Crush Zone could play in future, in the event of the re-emergence of Germany as the dominant political force in continental Europe. Fairgrieve, (1915) wrote:

> In touch with the sea and tempted on the ocean, [Germany] is one of the sea powers, while her position on the western and most populous margin of the great heartland makes her, at any rate, a possible centre from which that heartland might be organized.

It required no great imagination oh his part to picture the consequences for Britain, if Germany were to gain control over the heartland. Like Mackinder (1919), he also advocated that Britain should remain alive to such a possibility, and she should try her best to avert any such eventuality by forging unity between the countries of the maritime world under her own leadership, and by preventing the rise of any close alliance between Germany and Russia (Fig. 2.7).

Spykman and the Rimland Theory

In 1942, Mackinder's Democratic ideals and Reality was reprinted in the United States in its original form with a forceful introduction by the late Edward M. Earle of the Institute of Advanced Study in Princeton. Earle wrote: "there is no better statement anywhere of the facts of geography which condition the destiny of our world", and concluded his introduction by stating that the book was an essential reading for anybody interested in the study of strategy and international conflicts. This enthusiastic presentation of Mackinder's 1919 book without a word of criticism evoked a spirited response from Professor Nicholas J. Spykman (1893-1943) of the University of Yale, in a book published posthumously in 1944.

Working from the same premises as Mackinder, Spykman gave a very different interpretation of the relative importance of the Heartland vis-a-vis the surrounding tier, the Inner or Marginal Crescent, "partly continental and partly oceanic", which Spykman renamed as the "Rimland" (Fig. 2.8).

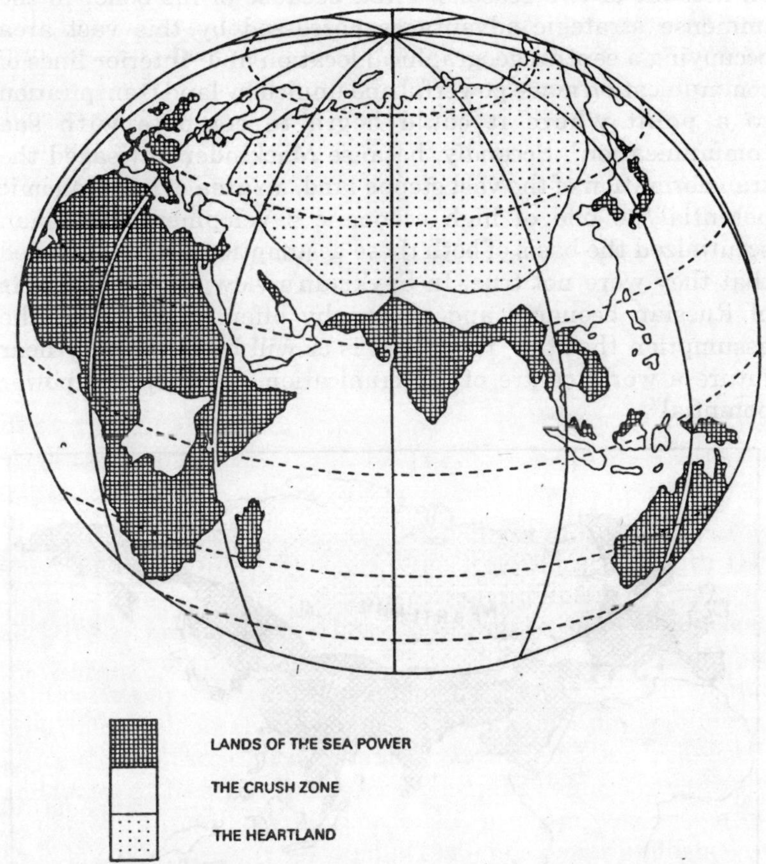

Fig. 2.8 : Lands of the Sea Power; The Crush Zone; and The Heartland (After Fairgrieve, 1915).

Drawing attention to maritime mobility as the basis of a new type of geopolitical structure, the overseas empire, in contrast to the former pattern of great land powers based on the control of contiguous land masses, Spykman emphasized that "It is sea power which has made it possible to conceive of the Eurasian continent as a unit and it is sea power which governs the relationships between the Old and the New Worlds". To him sea power appeared as the key factor in global strategy.

Spykman underlined that the great importance that Mackinder had attached to the Heartland region was chiefly on account of two reasons. First, because of his belief in the immense strategic advantages possessed by this vast area occupying a central geographical location and "interior lines of communication made powerful and unified by land transpiration to a point where it could begin to compete with sea communication". Secondly, because Mackinder envisaged the transformation of the vast steppe land, an area of low economic potential, to one of high economic development. Spykman scrutinized the basis of both these assumptions, and concluded that they were not true. In Spykman's view, the actual facts of Russian economy and geography offer no basis for the assumption that the "Heartland is or will be in the very near future a world centre of communication, mobility and power potential".

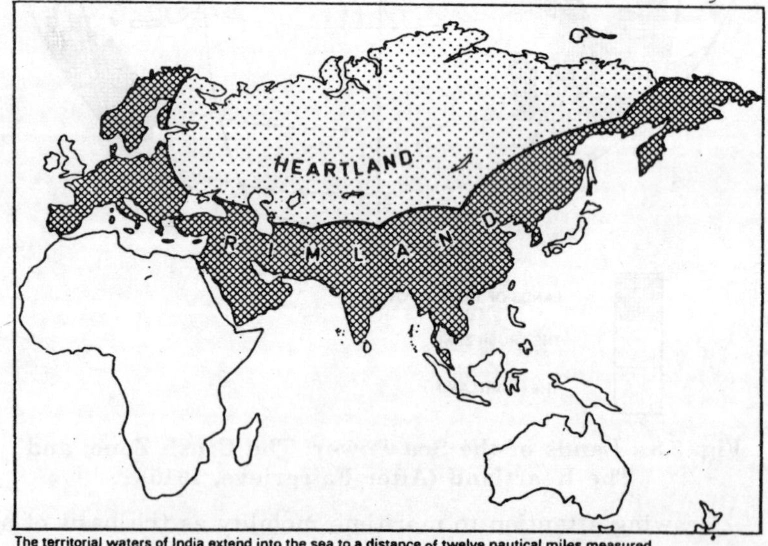

The territorial waters of India extend into the sea to a distance of twelve nautical miles measured from the appropriate base line

Fig. 2.9 : Spykman's concept of the Rimland.

In his opinion, there were many reasons to support such a view. In the first place, the distribution of world climate made it certain that unless revolutionary changes in agricultural technology took place, the centre of agrarian productivity shall continue to be located in Western Russia rather than in the central Siberian region. In this region the actual extent of arable land was, only a very small part of the total area. The essential bases for industrialization, at that time, also were mostly found to the west of the Ural mountains. Spykman conceded that the position in this regard had considerably changed, but he remained convinced that the Heartland "is not capable of supporting a large population from the produce of land". Furthermore, Spykman thought that despite the increased mobility created by the railways, automobiles and the aircraft in the centre of the Eurasian land mass, it could not be ignored that

> this area is ringed to the north, east, south, and southwest by some of the greatest obstacles to transportation in the world. Ice and freezing temperatures for a large part of the year, and towering mountains pierced by only a few difficult passes, form its borders. A large part of the rimland areas which touch the heartland/have even poorer transportation facilities.... The law of inverse ratio of power to distance remains valid within the same political unit as well as between political units. Within the immediate future. Central Asia will undoubtedly remain a region with a fairly low power potential.

Evaluating Mackinder's views regarding strategic advantages accruing to the Heartland from its position in the centre of the Eurasian land mass—a position which conferred upon her the advantage of interior communication with the countries of the rimland—Spykman wrote:

> It is obvious that the problems of an army which is working along the diameters of a circle or territory will be less difficult than those of forces which have to function along the circumference of the same region. In comparison with the exterior lines of British naval power running from Great Britain through the circumferential highway around the Eurasian rimlands, Russia has interior lines of communication...It must be pointed out, however, that interior lines function in terms of two points of reference rather than one. The

> relations between the centre and the circumference may easily be changed if a point on the circumference becomes in turn the centre of another circle of communication. Thus the strategic implications of the position of the heartland in relation to the British Empire have meaning only if the military strength to be applied at the Indian frontier originates in Great Britain. The moment the defence of that frontier or the Persian or the Chinese frontier rests on a locally developed war potential, the whole concept of interior and exterior lines is changed.

Spykman also criticized Mackinder's blanket application of the theory that the Old World history is a continuing struggle between land power and sea power. He wrote: "there has never really been a simple land power-sea power opposition. The historical alignment has always been in terms of some members of the rimland with Great Britain against some members of the rimland against Russia, or Great Britain and Russia together against a dominating rimland power". To Spykman, therefore, Mackinder's dictum about global strategy, appeared to be false. He wrote that if at all, there is need for a slogan for global power politics, it must be:

Who controls the rimland rules Eurasia;

Who rules Eurasia controls the destinies of the World.

To Spykman the Heartland appeared "less important than the rimland", and he was convinced that it is a combination of land and sea powers controlling the rimland, that would in all probability control "the essential power relations of the world"; inevitably so, because this combination would emerge as the owner of far greater resources, man power, and mobility, than the Heartland, or any other combination.

Writing at a time when the Second World War was still on, Spykman advocated that the Allied powers should base their future policy on preventing any consolidation of the Rimland by the enemy. With the defeat of Germany in the Second World War and the emergence of the USSR as the sole master of the Heartland, Spykman's prescription became the basis, at least in part, of the American policy of "containment" of communism.

It cannot definitely be said that the American policy of containment as advocated by George F. Kennan or that pursued by Truman, was in fact influenced by Spykman. However, ever since the War, very much in line with Spykman's prescription, the United States had consistently tried to build a tier of defence against the Soviet Union.

The different military alliances and treaties with the rimland States in Europe and Asia have been a part of this American policy of "containment"—the policy of surrounding and sealing the Heartland fortress. It also lies at the root of the socalled "Domino Theory" (which held that containment was necessary because the fall of any single country in the Rimland to the Soviets would have inevitably led to defeat of American interests in other adjacent countries), as well as the more recent concept of Finlandization (which maintained that the extension of the Russian influence in the Rimland need not have involved any direct take-over. Influence could more easily be extended through the control of the domestic policies of the relevant bordering States, as in the classic case of Finland).

Haushofer and the German World View

As noted in chapter 1, one of the basic ideas behind the theoretical frame of German geopolitics was Mackinder's theory regarding the heartland. Haushofer was keenly interested in geopolitical constructions derived from the arrangement of land and water bodies on the earth surface. Mahan had been a highly respected figure in imperial Germany and his ideas on sea power had been carefully studied. Haushofer himself was greatly attracted by Mackinder's 1904 paper. In his estimate, Mackinder's pivot-area thesis was "the greatest of all geographical world views". Haushofer wrote that he had never seen "any thing greater than these few pages of geopolitical masterwork" (Haushofer, 1936). His wife, Martha Haushofer had translated Fairgrieve's Geography of World Power into German, which was published (with an introduction by Karl Haushofer) in 1925.

The Geopolitiker's love for the heartland theory is easily explained in view of his, deep faith in the theory of the State as an expanding spatial organism. Mackinder's formula "Who rules East Europe commands the Heartland" appeared to offer Germany the much needed clue to achieving the objective of larger living space. One of the guiding principles of German policy in the post-1919 period was to gain control over Eastern Europe, where the Treaty of Versailles had erected a barrier consisting of several small states, with a view to bar Germany's expansion in that direction. It was a major objective of Germany's eastern policy to cut through this barrier and to incorporate these areas under a German dominated system. Once this was accomplished, the Heartland could be slowly penetrated through persuasion and friendship. Other logical steps could follow leading to a German dominated world system. This was to be accomplished through a treaty of mutual friendship with the Soviet Union. Since in the post-1917 period the Soviet Union, like Germany herself, was generally rejected by the West as an undesirable State (one not worth maintaining friendly relationship with) the two powers could themselves form, in the words of Haushofer, the "continental union of the expelled against the sea-ruling world powers".

Haushofer put his own interpretation on the original world map of Mackinder, in so far as the German nationalist world view differed from the one visualized by Mackinder from his location in London, and focused on the promotion of politico-economic interests of Britain. Mackinder's map showing the three-tier arrangement of land areas on the earth's surface, with the pivot area in the centre, was redesigned to project the German view. Two innovations introduced into this map, with a view to project the German world view, were:

(1) Depiction of an inner Hemisphere enclosing areas of higher civilization (including the German realm), and
(2) Identification of an outer Hemisphere dominated by centres of maritime power under British leadership, and trying to squeeze the German cultural realm of the Inner Hemisphere out of existence. Large areas in the Inner Hemisphere in South and South-east Asia, and the Middle East (*i.e.*, south-west Asia) were already under British domination. (Fig. 2.10) From the German point of view, it was necessary to break the maritime strangle-hold.

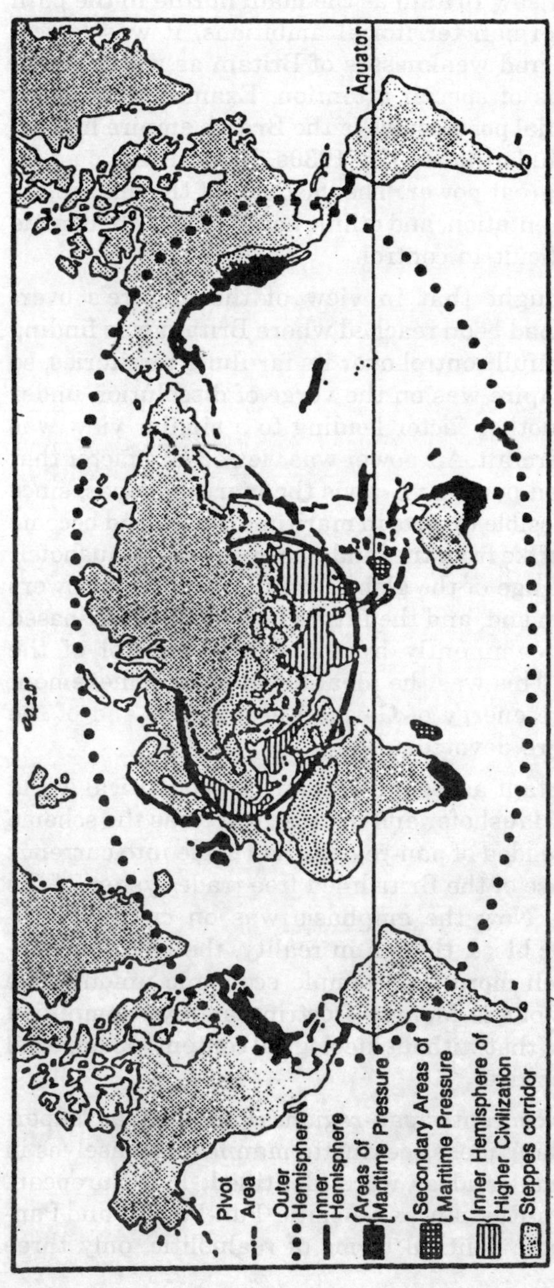

Source: Haushofer, K., Jenseits Der Grossmachte (B.G. Teubner, Leipzing, 1932).

Fig. 2.10 : Mackinder's map as adopted by Haushofer, (Source: Haushofer, K., Jenseits Des Grossmiichte B.G. Teubner, Leipzing 1932).

Since Germany saw Britain as the main hurdle in the path of realizing the German territorial ambitions, it was logical that the strengths and weaknesses of Britain as world power should be the focus of special attention. Examination of the structure and colonial possessions of the British empire had led Haushofer to conclude that in the 1930s, Britain was past its historic glory as a great power. The vast size of the empire, its geographical fragmentation, and ethnic heterogeneity had made it increasingly difficult to control.

Haushofer thought that in view of the empire's over-extension, a stage had been reached where Britain was finding difficult to exercise full control over its far-flung territories, so that the British empire was on the verge of dissolution under its own weight. Another factor leading to a similar view was the arrival of the aircraft. Air power was viewed as a factor that favoured land-based powers vis-a-vis the maritime ones, since the hitherto inaccessible targets in maritime States had become vulnerable to air strike from the land-based powers. Haushofer, concluded that the age of the global domination of sea powers had now come to an end, and the future lay with a broad-based land power—most eminently by the one in control of the Heartland region. This was the ideal toward the achievement of which the entire energy of German geopolitics, and of the German State, were devoted.

Another important aspect of Germany's geopolitical world view promoted by Haushofer and his Institute, was the scheme of pan-regions. The idea of pan-regions had come into currency following the demise of the British-led free trade system of the mid-19th century. Now the emphasis was on creating self-sufficient economic blocs, though in reality, the blocs (or pan-regions) were much more than simple economic unions. The Pan-Americanism of the Monroe Doctrine amply exemplified it. Haushofer wrote that without ideological content, no imperial structure could survive.

He defined his concept of pan-regions (Panideen) as "super-national all-englobing ideas seeking to manifest themselves in space". Six such major ideas were identified: Pan-European, Pan-American, Pan-Russian, pan-Asiatic, Pan-Pacific, and Pan-Islamic. In economic political terms of realpolitic, only three large pan-regions—Pan-America led by the United States, Greater East Asia led by Japan, and Pan-Europe to be led by Germany—were identified. Though as Parker (1985, p. 73) noted, in the 1941 version (produced before the German invasion of the Soviet Union, and while the German-Soviet pact was still in operation) Pan-Europe stopped at the Russian frontier, but incorporated the Mediterranean basin, the Middle East, and the whole of Africa.

Since the pan-regions were longitudinally aligned, they contained areas falling under all the three major climatic zones—a factor that ensured each of them a sound resource base providing for greater self-sufficiency. It was envisaged that this tripartite economic-political regionalization should, for a time, provide a kind of global balance of power (until of course Germany was able to work out the heartland strategy of Mackinder to its logical end of a new global empire, dominated by herself). The Second World War and the subsequent dominance of the world economy by the United States of America, brought the pan-region concept to an end. (Fig. 2.11).

Round World View of the Heartland Theory

As Bowman put it, by the end of 1930s "Geopolitics had migrated from Germany to America" in the sense that now the atmosphere of isolationism that, in the past, had led to "a strange provincialism" in the American geopolitical perspectives (Spykman, 1944), had been rapidly rectified, largely through the efforts of American political geographers like Hans Weigert (who was perhaps the most important contributor). With the help of imaginative maps, using new techniques and unfamiliar projections, these geographers enlightened their countrymen about the salient features of contemporary world politics, and the geographical reality against which the global actors were performing.

As Weigert put it, this task entailed sweeping away of "yesterday's geography", built around the flat earth Mercator projection, and letting the people "see the world in the mirror of dynamic maps, instead of the static maps of the times past" (Weigert, 1942). This was the beginning of an alternative geopolitical world view as seen from the other side of the Atlantic.

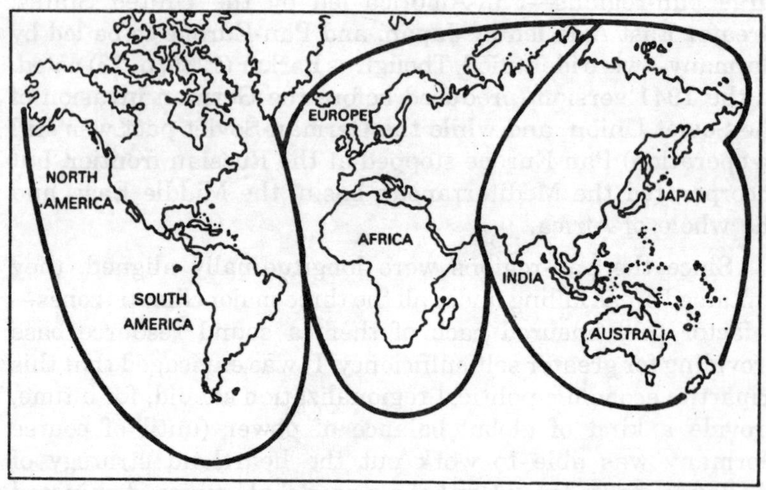

Fig. 2.11 : The pan ideas in 1941.

Weigert underlined that the real world was not flat but round, and that the flat earth map used by Mackinder to illustrate his pivot-area (heartland) thesis had led to a false sense of world geography and, hence, of global relationships between the major centres of power. (Mackinder's flat-earth map had also contributed to the deep sense of isolationism in the American mind, since by depicting America on the sidelines, and placing Europe at the centre, the Mercator projection map had fostered the impression that geography did not favour a central role for the United States in world affairs.)

Weigert, who was, personally, a great admirer of Mackinder's heartland thesis, reinterpreted Mackinder by representing his strategic scheme on an azimuthal equidistant projection centred in the Pivot-Area. In a joint paper, Harrison and Weigert showed that although enclosed in an elliptical outline, Mackinder's map was actually drawn on the Mercator projection. The two authors extended Mackinder's oval map out to the rectangle of the Mercator projection (Fig. 2.12), and demonstrated that Mackinder's 1904 map gave a more realistic picture of the relationship of the Heartland to other regions when transferred from the Mercator to the azimuthal equidistant projection (Fig. 2.13).

As Weigert wrote, it was impossible to visualize the Heartland as surrounded by an Inner and Outer crescent in the manner depicted by Mackinder, if the Mercator map was replaced by one drawn on an azimuthal equidistant projection. On an azimuthal

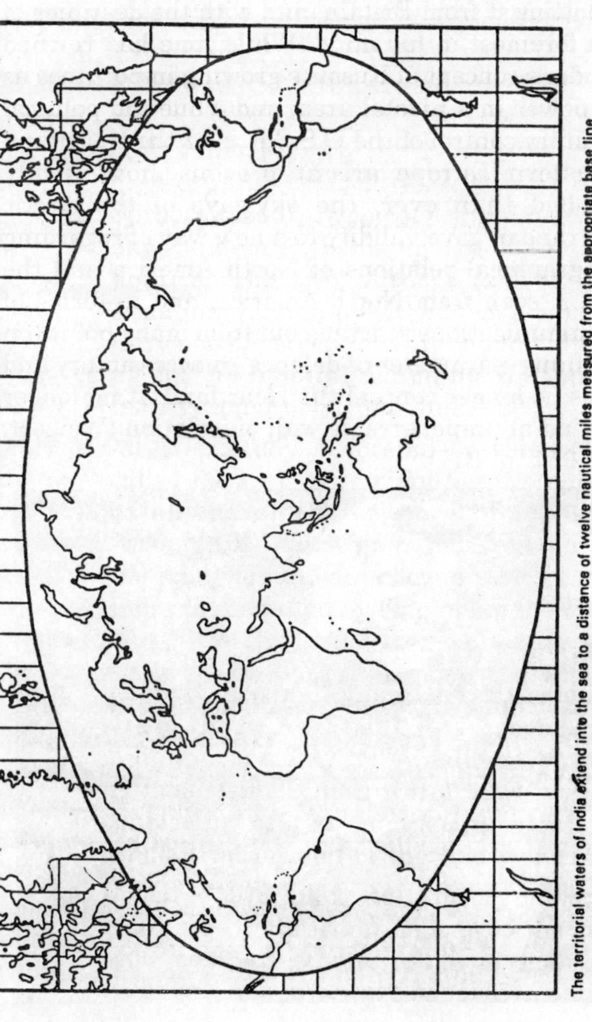

The territorial waters of India extend into the sea to a distance of twelve nautical miles measured from the appropriate base line

Fig. 2.12 : Mackinder's 1904 Map unveiled : Although Mackinder's 1904 map of the Heartland was drawn on the Mercator projection (on which areas in the higher latitudes are shown much larger than their actual size), by enclosing the map in an oval outline we created the misleading notion as if the map was drawn on some equal area projection. The Mackinder projection had helped to create a highly exaggerated image of the vastness of the Heartland. (Source: After Weigert et al., 1944).

projection (as on the globe) North America and the Heartland stand face to face across the Atlantic, so that,

> In terms of air-geography the Heartland and North America appear in destiny-laden proximity. As viewed over the top of the world, the Heartland assumes a location different from that which Mackinder assigned to it, plotting it from Britain, and with the destinies of Britain foremost in his mind. While time has verified Mackinder's concept of Russia's growing importance as a land power in a pivotal area, and while the political and military control of the U.S.S.R. over the Heartland and Eastern Europe are at present more firmly established than ever, the skyways of the Arctic Mediterranean gave validity to a new way of regarding the geographical relations of North America and the U.S.S.R.... seen from North America, and in terms of new communications reaching out from many points on the far-flung 'parameter of defence', inaccessibility and vastness no longer conceal the Heartland. It no longer lies behind an impenetrable wall of isolation (Weigert, et al., 1957, p. 217).

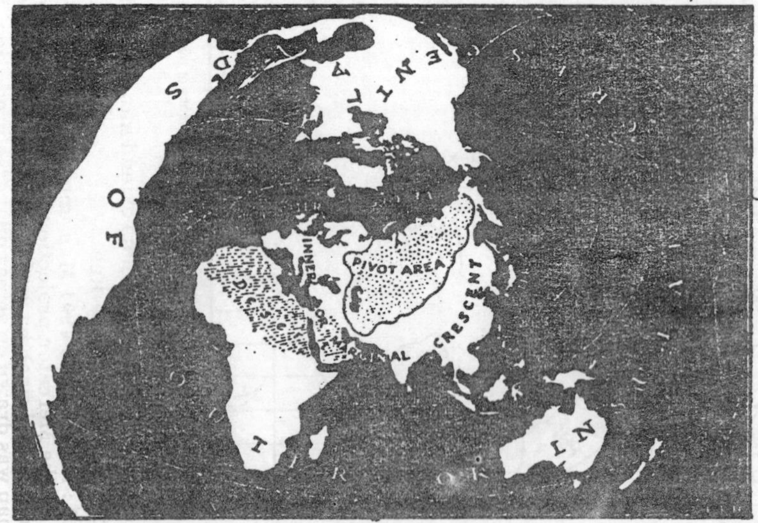

Fig. 2.13 Mackinder's 1904 Map transferred to an azimuthal equidistant projection centred on the pivot area (Source: Harrison and Weigert, 1943.)

MEINIG AND THE FUNCTIONAL CRITERIA

The Heartland-Rimland debate had received the attention of a number of political geographers (East, 1950 and 1956; Hall, 1955; Mills, 1956), but a paper by Donald Meinig (1956) made a somewhat more significant contribution, in that it drew attention to the dynamic and functional aspects of the Heartland-Rimland relationship. Meinig stressed the point that the difficulty with both Mackinder and Spykman lay in the fact that each wrote from a limited perspective of the contemporary power alignment of his time.

This, he noted, led to inflexibility of their strategic concepts, and, thereby, rendered them irrelevant to ground realities, following later changes in communication technology. Meinig maintained that in order that the Heartland-Rimland concepts can be made "applicable beyond the momentary context of strategic patterns", the two concepts should be so defined that they can be useful in the frequently changing historical-spatial situations. According to him, "our definitions of heartland and rimland must...be rooted in cultural or...functional criteria." He pleaded for a functionally defined concept of the Heartland based on broad similarity of physical conditions, basically similar cultures, and an interior nodal position possessing the nexus of historic land routes interconnecting various ancient centres of civilization on the oceanic margins of Eurasia. This contrasted with Mackinder's primarily hydrographic definition of the Heartland,

Of far greater practical value is Meinig's functional re-evaluation of the Rimland. He criticized both Mackinder and Spykman for regarding this entire crescent as a natural domain of sea powers of the world. Meinig thought that this crescent should more realistically be viewed in terms of its political rather than physico-geographical orientation in relation to the Heartland, and its rival combination of sea powers, since it is the political alignment in respect of the rival blocs that is functionally important. Since actual functional/political orientations of the individual States in the Rimland has kept on changing, in response to changes in the global power balance, it is wrong to assign the whole of this surrounding tier of States, a sea power orientation. The case of Yugoslavia, for instance was most revealing. It was a socialistic State like the erstwhile Soviet Union, and had for sometime been oriented towards the Heartland. But, later on, it assumed a functionally neutral political posture.

Likewise, Pakistan had gradually changed from a most pronounced American orientation in foreign relations, to a somewhat more neutralist, or transitional stance in later years. Meinig suggested, that on a functional basis, the Rimland should be viewed as divided into an inner (*i.e.*, towards the Heartland) oriented and an outer (*i.e.*, towards forces in opposition to the Heartland) oriented rimland. He termed the former as continental rimland and the latter as maritime rimland. Meinig suggested that whenever global politics was to be viewed in terms of the Mackinder thesis, the dynamic criterion of functional orientation within the Rimland must be kept in view. However, the collapse of Soviet union and the end of cold war have rendered this distinction to be of little contemporary relevance.

THE COLD WAR RELEVANCE OF THE HEARTLAND THESIS, 1945-1991

As Pearce (1962) wrote, at the first thought, Mackinder's work might seem superseded, just as developments in continental communications had enabled Mackinder to supersede Mahan. "But since the American and Soviet governments had committed themselves—by action and by words-to an indefinite stalemate in strategic weapons, the direct conflicts in the cold war had been waged by carefully limited land, sea and air forces in the peninsulas of the World Island. Par from being outmoded [Mackinder's] *Democratic Ideals and Reality* appears to be more relevant than ever" (Pearce, 1962, preface). In human affairs it is not "reality" but the "images of reality" that really count. War lives in the minds of men, and so does all politics. The Second World War which finished Germany as a military power, and severely crippled the rest of Europe, also saw the rise of the Soviet Union as the strongest power in Eurasia. Since the Soviet Union was a socialistic country, this gave a new clout to the ideological struggle between Communism and the capitalistic countries of the West. In view of the strategic advantages enjoyed by it, by virtue of its control over the Heartland, the Soviet Union soon began to be looked upon, in terms of the Mackinder thesis, as a colossus controlling the fortress, and threatening to sweep over the rest of the world. This was particularly so because the post-Second World War period saw a rapid expansion of Communism in areas along the periphery of the Heartland. Besides the East European countries, China had also gone the Communist way. The West began to picture the entire Communist world as a closely organized bloc acting in unison.

The pre-1991 cold war psychology of the non-Communist West, was aptly summed up in the following statement.

> The geographical arrangement of the territorial bases of both the major contestants suggests that the struggle is likely to take the shape of an onward pressure by the members of the Communist bloc and of the attempts by the encircling tier of States to restrain this pressure. If this is so, then the analyses presented by Mackinder, Spykman, and others have a very considerable element of truth and may be said to give us a framework within which to analyze the international situation. We are being confronted with a series of outward jabs from the bloc countries, directed now toward Berlin and Germany, now toward Turkey, Iran, or Afgahanistan, now toward Laos, Vietnam, the offshore island of Formosa, or Korea. The pressure is not always long sustained, and it is not in every instance military, though it is generally backed by the armed strength of the Soviet Union or, to a lesser extent, of Mainland China. These events arc suggestive of that "persistence of geographical relationship" to which Mackinder referred, and they appear to justify some kind of policy of containment....
>
> We dare not assume that old-style containment thesis has outlived its usefulness. Expansion from the communist States of Eurasia by traditional means of aggression or attack is easier, cheaper, and less likely to provoke violent retaliation than the use of the newest rocket-propelled nuclear warheads. Any weakening of the ability of the West to bring traditional kinds of

> military aid to the containing nations, such as Turkey, Iran, Pakistan, or South Korea, would invite attack on these countries with traditional weapons. Any loss of momentum in the build up of nuclear weapons might invite a kind of atomic blackmail by the Soviet Union or China.
>
> There are two alternative patterns of strategy, the traditional and the nuclear, and one of the strongest weapons in the Communist armory is its apparent ability to dictate the technological pattern of a future war and to prepare for it. The non-Communist world has to plan for the possibility of war at two levels, localized, brush-fire wars, roughly conforming to the geographical pattern prescribed by Mackinder, and also global and nuclear war. Containment is no longer enough nor is it possible today.
>
> On balance, it seems likely that the Soviet Union and China may in turn knock on the doors of the enclosing and containing tier of States. This was the weapon of the Russian czars and even of their predecessors, the Mongol tribes. It continues to be a weapon of the Soviets, supplemented and reinforced by the threat of nuclear attack in depth and by the opportunities, wherever and whenever they arise, of revolution far behind the lines. A three fold policy, containment, the nuclear deterrent, and economic aid, is necessary to withstand this three-pronged strategy (Pounds, 1972, pp. 436-437).

Whatever its shortcomings—and there are many—the Heartland theory continued to be relevant to the post-1945 bipolar world politics, because the parties to the game of power continued to have faith in it. Deepseated faith in the Heartland theory was at the root of the policy of nuclear deterrence. Once it was accepted that the heartland endowed the erstwhile Soviet Union with geopolitical superiority vis-a-vis the West, it became necessary that the West should go in for massive arms build up—a nuclear arsenal—in order to counter-balance the strategic advantages possessed by the Soviet Union. (Mackinder's relevance in the changed contexts of 1990s is discussed on pp. 134–138.)

THE ROUND-WORLD PERSPECTIVE : GLOBAL STRATEGY IN AN AIR AGE

Mahan, Mackinder, and Spykman speculated strategy essentially in terms of a global struggle between land and sea powers. In that context the Ksartland did appear as a huge fortress, irrespective of whether one favoured land or sea power as the more important of the two. But after the First World War, the destructive potential of air bombing was clearly demonstrated, so that any scheme of global strategy that failed to take air mobility into account, was in that measure unrealistic. Writing in 1943, Mackinder remarked that "some persons seem to dream of global air power which will 'liquidate" both fleets and armies".

He himself remained convinced, however, that "no adequate proof has yet been presented that air fighting will not follow the long history of all kinds of warfare by presenting alternations of offensive and defensive tactical superiority, meanwhile affecting few permanent changes in strategical conditions". In the very same year as Mackinder madc this statement, Edward Warner noted that, "the fundamental doctrine of air power supremacy" is that the airplane possesses such ubiquity, and such advantages of speed and elevation, as to possess the power of destroying all surface installations and instruments, ashore or afloat, while itself remaining comparatively safe from any effective reprisal from the ground". As Warner further wrote,

> If the correctness of that doctrine be conceded, the predominance of the role of air force in military establishment—and the necessity of planning all campaigns with primary reference to creating the most advantageous conditions for one's own air forces and the most unsatisfactory for the enemy's—obviously and automatically follow (Warner, 1943, p. 485).

General Giulio Douhet (1869-1930) of Italy and General William Mitchel of the United States were, by the 1920s, already advocating the supreme necessity of air power, in order to meet the challenge of the emerging era of "total war". "The prevailing forms of social organization have given war a character of national totality—that is, the entire population and all the resources of a nation are sucked into the maw of war", wrote

Douhet, and added that "future wars will be total in character and scope" (Douhet, 1942, pp. 47-48). The American air power strategist, Mitchel, wrote that "future wars will again be conducted by a special class, the air force, as it was by the armoured knights in the Middle Ages.

Again, the whole population will not have to be called out in the event of a national emergency, but only enough of it to man the machines that are the most potent in national defence". He thought that in order to "gain a lasting victory in war, the hostile nation's power to make war must be destroyed—this means the factories, the means of communication, the food producers, even the farms, the fuel and oil supplies, and the places where people live and carry on their daily lives" (Mitchel, 1925, pp. 19 and 126). "The advent of air power, which can go straight to the vital centres and either neutralize or destroy them, has put a completely new complexion on the old system of making war. It is now realized that the hostile main army in the field is a false objective, and the real objectives are the vital centres" (Mitchel, 1931, p. 255).

Strategic Ideas of Alexander de Seversky

Ideas about the critical role of air power in global strategy, found a great exponent in Alexander de Seversky, who was greatly influenced by Mitchel, to whom his first book, *Victory Through Air Power* (1942), was dedicated. There were, however, two features of de Seversky's exposition that distinguished him from both Douhet and Mitchel.

First, he insisted on the vital importance of a large radius of action. Secondly, he "emancipated" the air force from the concern with extensive ground organization (that Douhet had considered necessary), and from dependence on the island stepping stones of the Arctic routes in the Atlantic and the Pacific (to which Mitchel had attached great importance). To De Seversky, maintaining advance bases, instead of hurling the full air strike potential directly against the enemy, appeared as a sheer wastage. He was convinced that "The entire logic of aerial warfare makes it certain that ultimately war in the skies will be conducted from home grounds, with everything in between turned into a no-man's land".

De Seversky's ideas were represented in his *Air Power: Key to Survival* (1950). In this book, in line with his earlier ideas, he advocated that the United States shorfid aim at the development of massive air power as its primary military/ defence objective. American superiority in air should be made so complete that no adversary might be tempted to engage in adventures against her, for fear of complete destruction at the hands of the American retaliatory forces. De Seversky insisted that overseas bases are costly luxuries, serving no useful purpose in offence or defence. The policy of containment was considered strategically outmoded in face of revolutionary changes in the art of war. Therefore, he advised American policy makers against involvement in small wars around the Russian periphery—such as the one in Vietnam—which only led to useless expenditure of valuable resources and energy, without doing any harm to the erstwhile Soviet Union.

De Seversky looked at the world as drawn on an azimuthal equidistant projection centred on the North Pole. Thus, he viewed the two chief adversaries for global supremacy as standing face to face across the Arctic Mediterranean, through whose air space, pass the shortest routes between places in the Heartland and North America. De Seversky's world map on the azimuthal equidistant projection led him to visualize both, the United States and the Heartland State, as separately holding complete dominance over vast areas enclosed within their respective orbits.

The two great circles marking their spheres of influence (and focusing respectively on the industrial heartlands of the USA and USSR) overlapped over a large area in Eurasia and North America (Fig. 2.14). As De Seversky pointed out, this zone of overlap was the area of critical importance in the struggle for global military supremacy between the two adversaries. He called this zone of overlap as the "area of decision". According to De Seversky, it was necessary for the United States, in the interests of its own security in the event of total war, that it should concentrate all its energies and resources in acquiring complete air dominance in this zone.

As Jones (1955) noted, the notion of absolute air supremacy based on intercontinental mobility through the air, and juxtaposition of the United States and the Soviet Union as the principal antagonists standing face to face across the Arctic, tended to encourage in the United States, a sense of "air isolationism" oriented to the Western hemisphere.

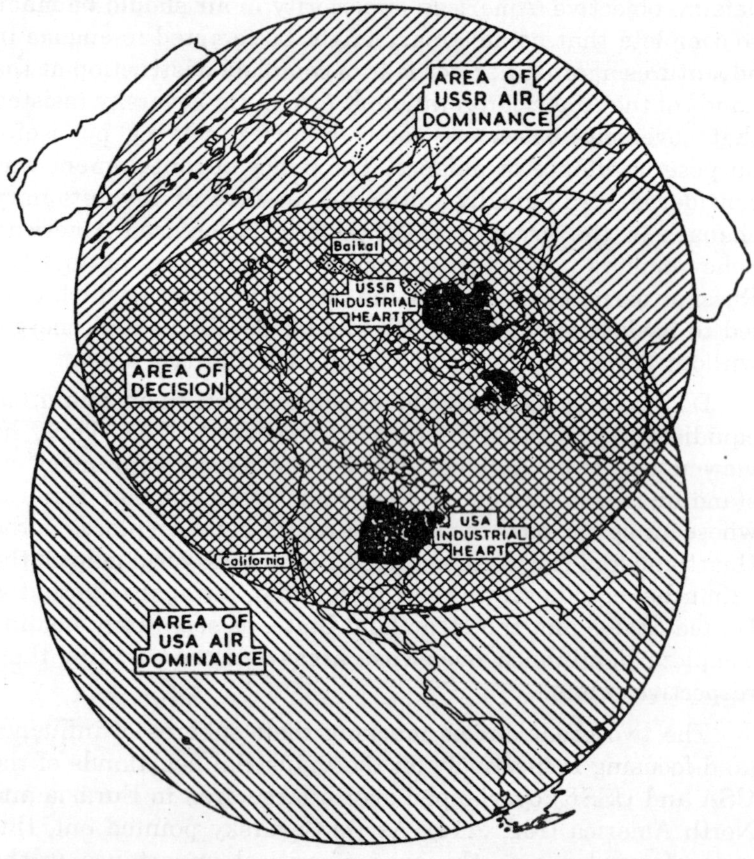

The territorial waters of India extend into the sea to a distance of twelve nautical miles measured from the appropriate base line

The territorial waters of India extend into the sea to a distance of twelve nautical miles measured from the appropriate base line

Fig. 2.14 : Airman's view of the world (After De Seversky, 1950).

De Seversky's generalization of global strategy also has not stood the test of time. From the perspective of the present, his views appear oversimplistic, keeping in view the further changes in transportation technology, and arrival of intercontinental missiles, nuclear-powered submarines, and satellites orbiting around the earth. In view of these revolutionary developments in warfare technology, De Seversky's analysis, concentrating upon air supremacy in relation only to the Arctic Mediterranean, has been rendered outmoded and unrealistic. Since De Seversky wrote, a further dimension to global strategy has been added with the increasing potentialities of sub-surface oceanic mobility based on nuclear-powered submarines. Any calculations about future strategy must take this new mobility into account.

A further difficulty with De Seversky's ideas was that he speculated about global strategy in terms of an all out war in a world neatly divided into two rival blocs, led respectively by the former USSR and the USA. But this is far from the ground reality. World politics is a multi-sector involvement, especially following the disintegration of erstwhile Soviet Union and the end of the Cold War. Any theory of global strategy must take all the different sectors and arenas into consideration.

International politics is a game not only of remaining one-up against one's adversaries, in terms of military superiority. It is, in fact, more truly a game aimed at enlarging the circle of friendly States through mutually beneficial economic and cultural relationships, and by enhancing the State's ability to stand effectively with friends and allies, in the event of limited wars. Global strategic interests of the United States could not, therefore, be served by her massive military strength in one sector only, as advocated by De Seversky. As Clausewitz said, the purpose of war is the pursuance of political control and influence by other means. This purpose could not be served by the De Seversky strategy, for "nuclear war could never be a continuation of politics" (Bundy, 1971).

It may be relevant here to note that the reason why nuclear arms became a necessary concomitant of air power was that: long range bombers would be ineffective weapons against the Heartland if armed only with chemical bombs, as flying such long distances would reduce the weight of the bombload. In addition, the efficacy of strategic bombing in the Second World War was by no means proven. Air power was not looked upon as the means of delivery for nuclear arms. The essential aspect of the deterrent was not air power per se, but nuclear weapons" (Walters, 1974, pp. 47-48).

STRATEGY IN THE AGE OF NUCLEAR POWERED SUBSURFACE SHIP

The Heartland theory has been the fundamental basis of the Western concept of nuclear deterrent. Nuclear arms seemed necessary in order to redress the supposed imbalance of power between the Soviet Union and the West, in terms of the Mackinder theory. But because total war can never be an instrument of policy, the excessive emphasis placed on air warfare and nuclear arms in Western strategy had greatly distorted their military thinking, with disastrous consequences to policy in national as well as overseas affairs. Secondly, under the influence of the of-repeated phrase, "air age," modern thinkers on strategy have generally tended to ignore the strategic role of seas, especially the new prospects of maritime power arising from the application of nuclear energy in the development of a new kind of ship—the undersea ship. As Hezlet wrote:

> The power of air craft over the submarine in the past has progressively been reduced as it developed into the "true" submarine. It is now only a matter of time before surface-to-air guided weapons are adopted for submarines and they acquire the power to shoot aircraft down....The slow anti-submarine aircraft and helicopter would be very vulnerable to a simple short-range missile launched from a submerged submarine...In the future, far from the submarine living in the fear of aircraft, it is likely to be the other way round, and aircraft flying over the sea will never be sure that a guided weapon is not about to be launched at them from an undetected-submarine (Hezlet, 1967, p. 261).

While the development of reconnaissance satellites has made the surface ship-nuclear-powered or conventional-an ineffective instrument of war by making the surface of the sea an open book; detection under water by sonar has lagged far behind the development of the undersea ship, and the range of detection is still measured in terms of thousands of yards. This is because,

> changes in salinity, presence of marine life, configuration of the bottom and, above all, layers of water of varying temperature that refract sound not only reduce ranges of detection but make the performance of detection systems erratic and unreliable. Moreover, the modem submarine is capable of perceiving and exploiting these anomalies so as to make the hunter's task as difficult as possible (Martin, 1967, p. 99).

As Walters noted, the undersea ship has destroyed two of the basic premises of Mackinder's theory. The conventional submarine which was viewed as a denial of sea power, has been relegated to a minor place, and rather than seeing the end of sea power, we are moving into a period which will bring a vast expansion of sea power. Secondly, the undersea ship by making the Arctic ocean navigable, explodes the myth of the Heartland's invincibility. It thereby makes the World Island truly an island that can now be freely circumnavigated. The opening of the Arctic Ocean for navigation lays bare new vistas for the future.

As Clausewitz wrote: "wars are in reality...only the manifestations of policy". Policy is "the intelligent faculty", the directing force, and "war only the instrument". The objective of war is the political structure, and the aims of war include the post-war settlement. And if war is to be viewed as an instrument of policy, a means to achieving political and economic objectives, then movement is of far greater importance than destruction. Viewed from this perspective, the nuclear-powered engine offers the greatest potential for mobility since it is the most modern form of propulsion. Nuclear-powered submarine has become, in the words of Roskill, "the capital ship of the nuclear era" (Roskill, 1970, p. 343). As Roskill put it,

> The scope for employment of submarines on duties hitherto performed by surface ships or aircraft appear to be limited only by the cost of submarine production. Indeed, the command of the surface sea, whereby a maritime power seeks to secure the undeterred passage of mercantile or military cargoes, seems likely to depend increasingly on control of the waters beneath". It is,

therefore, logical to conclude that "he who rules the undersea will control the surface", and although it is not obvious, "the undersea ship offers an alternative strategic doctrine to that of nuclear deterrent (Walters, 1974, p. 134).

When at a future date, the undersea ship becomes a viable commercial ship, the Arctic will become the strategic sea of the future with profound geopolitical implications. According to Walters:

> The Arctic Ocean would then be part of the narrow waters of the future in the same way as the English Channel has been of the narrow seas in modern times. The World Island is only an island in theory and not in practice, as Mackinder pointed out. If it were possible freely to use the polar Mediterranean, then the World Island would be an island in the true sense of the word. The navigation of the Polar Sea in 1958 by Commander William Anderson was a historic achievement of monumental geopolitical significance. Though largely ignored at the time, in the perspective of the future it may be seen to -be as important as was the rounding of the Cape of Good Hope by Vasco da Gama in 1497 {Walters, 1974, p. 149).

The undersea cargo ship of the future shall be able to take maximum advantage of the Arctic Mediterranean across which pass the shortest routes between the most industrialized nations of Europe and North America. Its importance will be further enhanced because lands around the Arctic—*e.g.*, Alaska and Northern Canada—are known to be rich in minerals, which remain largely unexploited owing to the economically prohibitive cost of transport.

THE FALSE OBJECTIVITY OF GLOBAL STRATEGIC VIEWS

Although the different global strategic views presented since the time of Mahan and Mackinder, were often projected as objective analyses of politico-geographic realities, each scheme, in its turn, represented a particular world view as seen by the concerned author from his home base, and with his country's current politico-economic interests supreme in his mind. In Mahan's case it was quite obvious. He played a leading role in persuading the United States to pursue a larger destiny overseas during the opening years of the present century (Sprout, 1963).

The politico-economic interests of Britain were supreme in the mind of Mackinder (1904), himself a practicing politician. At the time that Mackinder wrote, Britain's role in world economy was changing, so that it had begun to be realised that simple accumulation of capital in London, through a policy of free trade, was no longer sufficient to meet the challenge posed by Germany's massive growth in heavy industry. Tariff barriers and protectionist economic policy was urgently needed to maintain British industry, in face of the German challenge. These concerns, inevitably, found expression in Mackinder's world model (Semmel, 1960).

Mackinder's Heartland thesis was, in many ways, a concealed rationalization for the traditional British policy of balance of power in order that no European power could challenge her dominance. Mackinder's paper underlined that in the railway age, Britain was more vulnerable than ever before to an upset in the balance of power in Europe, so that it needed to prevent any alliance between Germany and Russia that could endanger British supremacy in the world politico-economic system. The central message of his 1919 book was that East Europe was the key to the Heartland, it required, therefore, to be converted into a buffer zone of small States between Germany and the Heartland.

Spykman's rimland theory, and the related idea of containment were clearly framed to project the interests of the United States as a super power. The same was also true of De Saversky's prescription for air power strategy, specifically designed to guide post-1945 American defence policy. As Cohen (1991) noted, "Objective" analysis cannot escape the experience and national biases of their authors. His own analysis, he wrote, "reflects an American point of view. Its prescriptions are directed to American policy-making".

GEOPOLITICS IN A SHRINKING WORLD

Emerging Realities of Space-Time Convergence in a Four Dimensional World

Revolutionary change in communications and transport technology, following the introduction of the aircraft, orbiting satellites, ICBM systems, and instant global communication via telephone and television, has greatly altered our perception of geographical space and its political organization. The emergence of the outer space and ocean depths as significant influences in global-political, economic and military manoeuvres, has transformed our world into a four-dimensional one. The greatly accelerated speed of transport and communication has resulted in vast reductions in the amount of time required to establish contacts between places, leading to a high degree of time-space convergence, rapidly converting our world-space into a "World of Points", wherein all locations (and distances between them) are relative. As Brunn (1981) noted, in an absolute sense, time-space convergence defines places that are equally easy to communicate with, irrespective of the surface distances between the sender and the receiver. These developments have helped greatly to reduce the role of international boundaries as barriers in the functional space, since in the present day world,

> No location can remain "closed" from over-expanding international airline networks, television cameras, and international ballistic missile systems. Everybody and everything is equally exposed to seeing-eye resource and military satellites, orbiting weapons, powerful radio transmitters, communications pirates and journalists who can report almost anything, from anywhere, at any time. (Brunn, 1981, p. 145).

The instant news coverage brought about by time-space convergence is bound to create information overload to which many governments in the less affluent regions shall find difficult to adjust. This is a potential source of "future shock" (that is, too much change in too short time, or a premature arrival of the future). It is rightly noted that in a political sense the political spaces and forms of government established during a period of little cultural interaction and political cooperation, as well as slow land transportation and little formal communication, become unwieldy with the diffusion of high speed technologies. Study of geopolitics in a rapidly shrinking world, therefore, stands out as a priority item of research in a futuristic mould.

GEOSTRATEGY IN THE AGE OF MULTIPLE POWER NODES

All through the post-war period until early 1960s, the several variations over the Heartland thesis were presented as modifications over the original framework, to explain the changed scenario of world politics. These attempts appeared fairly successful since the post-war emergence of a bi-polar world had contributed to reinforce the age-old contest between the Heartland and the surrounding maritime crescent. The picture had qualitatively changed by the end of the 1950s. The initially, tight bi-polar pattern of world politics between USA and USSR had started loosening up and the world was slowly moving towards a new world system comprising a number of regional political groupings forged with a view to safeguard shared regional interests of their constituents.

The tendency became particularly pronounced as a large number of new States joined the rank of sovereign nations, following the post-war withdrawal of European colonialism from Africa and Asia. A great majority of these countries—freshly released from the yoke of political slavery—were in no mood to curtail their freedom of action through complete absorption into the bi-polar system of super power politics. But since most of them were individually too small and weak to stand out on their own, they suffered from a general feeling of insecurity. This was sought to be offset through the strategy of regional cooperation and unity of action. In the 1960s, therefore, the world had entered the age of regional powers. Any realist theory of global politics had inevitably to address itself to the emerging regionalism in world politics.

A bold attempt at a realist geopolitical model, incorporating the regionalist ethos of the contemporary international system, was made by American geographer Saul B. Cohen in his widely referred *Geography and Politics in a World Divided* (1964,

1973). The basic purpose of his book was to question America's post-war policy of containment, and in that process "to present a framework that anticipates the geographic dynamism of our times". To begin with, Cohen rejected the basic premise of the containment policy—the supposed unity of space between the inner Eurasian regions (the Heartland), on the one hand, and the maritime Rimland on the other. He discounted any intrinsic strategic unity in the Rimland as a whole. He saw contemporary world politics functioning simultaneously at two levels of fragmentation. One was the fundamental division of the world into two-fold "geostrategic regions"—the Trade-Dependent Maritime World led by the United States on the one side, and the socialist bloc of States led by the USSR, on the other. Secondly, there was the lower order division of each of the two into "geopolitical regions".

The Trade-Dependent Maritime World was visualized to contain five "geopolitical regions". These are: Anglo-America and the Caribbean, South America, Maritime Europe and Maghreb, Offshore Asia and Oceania, and Africa-south of the Sahara. The Eurasian Continental World consisted of two geopolitical regions—first, the Heartland together with Eastern Europe, and second. East Asia. South Asia was described as a potential geopolitical region that might well .arise from the ashes of the British Commonwealth.

Between the two basic geostrategic regions were located two somewhat distinctive geopolitical regions—designated as "shatter-belts". These were. South-east Asia, and the Middle East. Unlike the other geopolitical regions, the shatter-belts lacked political unity and mutual sympathy, owing to which they have remained fragmented. As a result, both the geostrategic regions have footholds in them. The chief geopolitical significance of the shatter-belts lies in their strategic importance to both the super powers. This aspect contributed to make the two shatter-belt regions, the real testing ground in the contest for supremacy between the two super powers. Accordingly, it is in these two areas that "containment" was required to be pursued. Cohen advocated selective application of the policy of containment, according to the ground realities, rather than making it a blanket operation covering the whole of the Rimland. (Cohen's shatter-belts were functionally similar to Alexander de Seversky's "area of decision", the fundamental difference being that owing to Cohen's regionlist perspective, his scheme contained more than one area of decision.) (Fig. 2.15).

In an article published in 1982, Cohen modified his original scheme to designate Africa South of Sahara as the third shatter-belt. He drew attention to the fact that during the intervening period, three of his original geopolitical regions—Japan, China, and Maritime Europe—had developed into world powers standing alongside USA and the USSR.

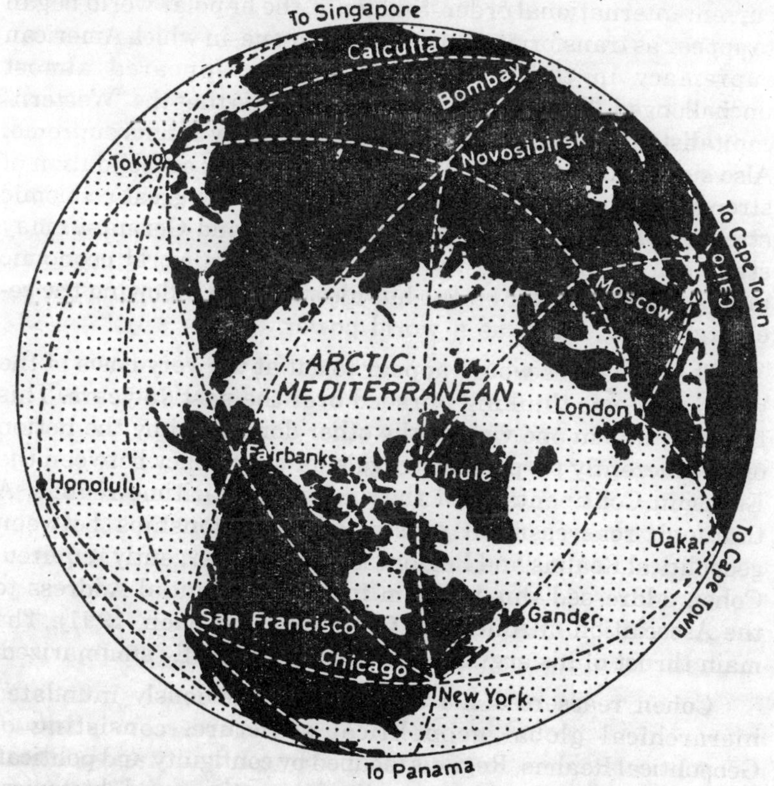

Fig. 2.15 : The Arctic Mediterranean.

Elsewhere, other geopolitical regions had developed into second-order powers dominating inter-State politics of their respective regions. Cohen included a total of 27 States in this category, the most conspicuous being India, Brazil, and Nigeria. Based on the scope for a country's influence beyond its own borders, Cohen identified a further hierarchy of third, fourth, and fifth-order States.

Global Geopolitical Change in the Post-Cold War Era

The traumatic changes in the world strategic scene following collapse of the Soviet Union, dissolution of the Communist Party, and independence of the Soviet satellites in Eastern europe, has completely changed the strategic complexion of the current international order. Suddenly, the bi-polar world began to appear as transformed into a uni-polar one, in which American supremacy in the international arena appeared almost unchallenged. The "East" has disappeared, so that the "Western" capitalist politics, economy, and ideology now reigns supreme. Also significant strategically, is the increasing consolidation of strength in the European Community, and the great economic strength of Japan to the point where she (and Germany) may stand side by side with the United States as world economic giants. China's power status has also declined following the re-emergence of Japan as a world power.

These changes would appear to cut at the very basis of the global strategic thinking pursued over the past 75 years. This thinking was in one way or the other dominated by the notion of an inherently perpetual struggle for supremacy between the Eurasian Continental and the encircling Maritime realms. A thorough reassessment of strategic relationships between geopolitical Realms and Regions is, therefore, urgently required. Cohen addressed this problem in his Presidential Address to the Association of American Geographers (Cohen, 1991). The main thrust of his argument may here be briefly summarized.

Cohen reasserted his belief in his previously inundated hierarchical global geopolitical structure, consisting of Geopolitical Realms, Regions (shaped by contiguity and political, cultural, military and economic interaction, and historical movement), Nation-States (hierarchically ordered according to their positions and functions in the world system), and Gateway Territories (embryonic States which can accelerate exchanges that may stimulate the evolution of larger nations, from which gateways have spun off). This last category is a new innovation in his scheme of global spatial structure (Fig. 2.16). Cohen pointed out that while most of the second level regions are contained within the larger realms, three lie outside them. These are: South Asia (which is now seen as an independent region), the Middle East (which is a Shatterbelt, caught between the two realms), and the emerging Gateways region of Central and Eastern Europe (a transitional zone that can facilitate contact and interchange between the two Realms). He anticipated the new emerging geopolitical phenomena—the

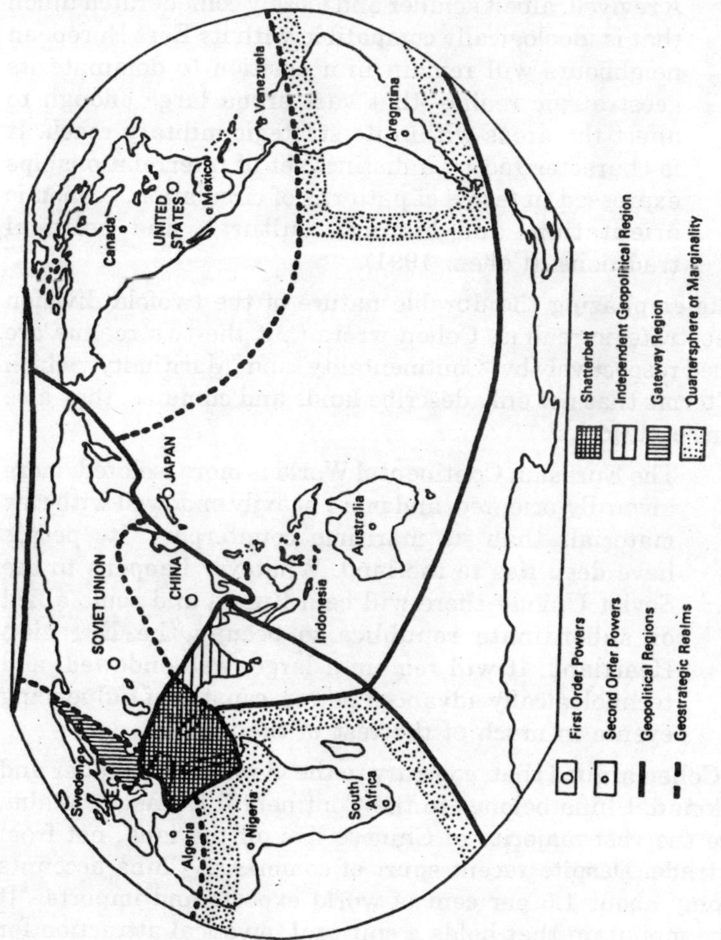

Fig. 2.16 : Cohen's world strategic realms and geopolitical regions in the 1990's.

Gateway region and Gateway States—to play significant roles in restoring balance to the world system.

Cohen noted that the deteriorating economic and political fortunes of the erstwhile Soviet Union may lead some to ask whether the concept of the Eurasian geostrategic realm still has validity, and argued that it is premature to dismiss the now reduced version of the former Soviet Union from the status of a controlling State in an arena of the world that has impact upon much of the rest:

> A revived, albeit smaller and loosely confederated union that is ideologically compatible with its East European neighbours will remain in a position to dominate its geostrategic realm—that vast arena large enough to affect the areas within its strategic military reach. It is characterized by a distinct set of interrelationships expressed in terms of patterns of circulation, economic orientation, and historic, cultural, and political traditions. (Cohen, 1991).

Re-emphazing the durable nature of the twofold division of geostrategic realms, Cohen wrote that the two realms are defined respectively by "Continentality" and "Maritimity", which are "terms that not only describe lands and climates: they also describe outlooks":

> The Eurasian Continental World is more isolated, more inwardly-oriented, and more heavily endowed with raw materials than its maritime counterpart. Its people have deep ties to the land. Whatever happens to the Soviet Union—there will be a Russia and some allied or subordinate republics to occupy the Eurasian Heartland. It will remain a large, well-endowed, and technologically advanced power, capable of influencing events in much of the rest of the world.

Cohen argued that, contrary to the views of Mackinder and Spykman, China belongs to the Continental Eurasian Realm, since the vast majority of Chinese live off the land, not from sea trade. Despite recent spurt of commerce, China accounts for only about 1.5 per cent of world exports and imports. "It is the mountain that holds a spiritual, mystical attraction for the Chinese, and not the sea. And it is the common border that strategically links U.S.S.R. [now Russia] and China. They cannot turn their backs on one another; they have to find a modus vivendi".

He defined Shatterbelts as regions that present "an equal playing field to two or more competing powers operating from different geostrategic realms". Thus, South Asia is not a shatterbelt since India's dominant position is not seriously threatened by the United States, Russia or China. On the other hand the Caribbean is under complete sway of the United States. Pointing to recurrent shifts in the character and positions of Shatterbelts, Cohen noted that Sub-Saharan Africa became a Shatterbelt in the 1970s, as Soviet satellites, such as Cuba and certain East European States, offered military aid to Ethiopia, Angola, Namibia and Mozambique; and the Soviet Union herself made political inroads into Congo, Guinea, Mali, and Tanzania. With the collapse of Soviet Union, Sub-Saharan Africa has now reverted back to the Maritime fold. South-east Asia has also lost its Shatterbelt character and its insular and southern peninsular portions have become economically and politically part of Offshore Asia and the Maritime World. Also with the fall of the Soviet Union, Vietnam and Indochina as a whole, are likely to fall within the East Asian sphere. As regards the isolated, impoverished Myanmar (Burma), after the repressive military regime is eventually over-thrown, it may, in all probability, become reoriented to South Asia. Thus, the only remaining Shatterbelt is the Middle East, which is itself in transition, and is increasingly tilting toward the Maritime Realm, as the Soviet Union suddenly ceased to be a major supplier of economic and military goods. The Middle East contains six regional powers. These are: Egypt, Iran, Iraq, Israel, Syria, and Turkey.

Toward Restoring Equilibrium between the Two Geostrategic Realms : Gateway Regions and Gateway States

Cohen argued that the world is currently in disequilibrium, owing to great differences in entropy levels between the two geostrategic realms. The argument runs that, owing to the closed nature of their political regimes, the Russian and Chinese cores of the Eurasian realm have brought their human and natural resource-base close to exhaustion. The closure of their systems to outside influences, led to denial of social, political, market, and technological energies that could promote innovation and renewal. According to him, equilibrium between the two realms could be restored only if their entropic levels are brought closer to equal. In Cohen's view:

> The most promising geopolitical mechanism now restoring the balance between the two realms is the emergence of Central and Eastern Europe as a Gateway region. Such a region could facilitate the transfer of new energies into the faltering Soviet core. Extending on the west from the Oder-Neisse rivers and Harz and Bohemian mountains to the Northern Adriatic Sea, and east to the borders of the [former] U.S.S.R., the European Gateway will be fully open to the economic forces from the east and west. Its national politics and economic structures are adopting the West European models, but it will have to find a military posture that does not challenge Soviet [*i.e.*, Russian] security goals. With the exception of Greece, this region is composed of that middle tier of States between Germany and Russia whose independence and stability Mackinder (1919) felt to be crucial to Eurasian and world stability.

Cohen argued that the reason why East and Central Europe will become a Gateway Region, rather than continue as a Shatterbelt, is that both Russia and the European Community shall find competition as counterproductive. Maritime Europe's concerns are the Russian military power, and Russia needs West European economic help. The two sets of needs balance one another. Cooperation rather than competition shall fulfil these needs.

Cohen speculated the possibility of the Caribbean and Central America emerging as a future Gateway region. The Caribbean is at present, completely under United State's sway (the diminutive Cuba notwithstanding). The disappearance of Russian influence from this region gives scope for such regional powers as Mexico and Venezuela, to extend their influence since the United States shall now have little ground for worry on account of military issues. This way, the United States can rechannel its resources to regional development.

> With a native demographic slowdown and the growing difficulties of enlisting its poorly educated, drug-ridden underclass into the labour force, the U.S. draws upon the pool of labour from the lands to the south. In addition, Mexico represents a major focus for 'off-shore' American manufacturing, and the Caribbean basin as a whole has the potential for attracting Japanese capital as manufacturing points of entry to the American market.

Politically and culturally Gateway States represent distinct historic culture hearths. Each is distinguished by a separate language, and often separate religion, relatively higher degree of education, and favourable location for areas outside the State. Economically they tend to be more highly developed than the core areas of their host States.

Table 2.1 : Cohen's List of Prospective Gateway States.

By early 21st century		*By mid-21st century*
Catalonia	Madeira Islands	Alaska
Cyprus (united)	Mount Lebenon	Azores
East Nicaragua	Pashtunistan	British Columbia
Eritrea	Puerto Rico	Hawaii
Estonia	Punjab	Northern Ireland
Gaza/West Bank	Quebec	Northern Mexico
Hong Kong (Shenzhen)	Slovania	Soviet Far East
Kashmir	Tamil Eelem	West Australia
Latvia	Vascongadas (Basque)	
Lithuania		

Wherever they are regions of out-migration, they acquire links to groups overseas, that may become a source of capital flow and technological knowhow. Owing to their strategic location athwart key access routes, the Gateways are of considerable military value to their parent States. And despite their relatively advanced economies, the Gateways lack self-sufficiency, and depend upon their parent State for raw material and market.

Cohen thinks that the emergence of such "States" can facilitate the creation of boundaries of accommodation between the two geostrategic realms since now the conflict between them has abated. Table 2.1 lists his Prospective Gateway States.

MACKINDER AND THE GEOPOLITICAL SCENARIO OF 1990s

The current turbulence in the global political system, most particularly, collapse of the Soviet Union, end of the Warsaw Pact resulting in independence of the Soviet satellites in Eastern Europe, followed by armed American intervention in the Gulf, with moral and material support of her European allies, and tacit approval of such an action on the part of the Russia and the Eurasian Continental realm, presents a geopolitical scenario that appears to indicate that contemporary international politics may well have bidden the final good-bye to the so far highly influential Heartland theory of Sir Halford Mackinder (1904, 1919), as also to the foreign policy formulation based on his notion of a perpetual tug-of-war between the Continental Eurasian realm and the encircling crescent of Maritime States.

The world of international politics is very complex, however. What is apparent may not necessarily be real. The emerging geopolitical structure therefore needs to be closely analyzed in order to ascertain the correct picture. This is the purpose and objective of the paragraphs that follow.

The collapse of the Soviet Union and the end of four decades of Cold War, has transformed the bi-polar geopolitical structure of the world into a unipolar one. In the opinion of Krahthmmer (1991), "The most striking feature of the post cold war world is its Unipolarity". Super-predominance of the United States in world affairs is unquestioned. Serious doubts do, however, exist regarding the precise nature of the transformation of contemporary world order. In the first place, military power is no longer the sole measure for determining power position in the comity of nations, much less the status of any State as a super power. Thus, the collapse of the Soviet Union need not necessarily lead to American overlordship of the world (Pfaff, 1991). There is ground to believe that owing to its inherent multisectoral structure, world politics is bound to become, before long, a multipolar involvement. But the poles may have different currencies of power, *e.g.*, economic and financial (Japan and Germany), demographic (China and India), and military (United States and Russia)—and the nature of international politics shall, in all likelihood, depend on how the different poles cooperate with one another (Hoffman, 1990).

The collapse of the Soviet Union clearly signifies the loss of status of the Russian State as a super power, but it has by no means written off the Heartland region as a major player in international politics. Whatever form the political formation replacing the Soviet union (presently designated as the Commonwealth of Independent States—CIS) may take, and howsoever reduced in size and strength it may become, by virtue of its geographical location, its resource endowment, and economic and military power (which shall continue to be greater than that of any State, other than the united States), it is destined to remain a leading force in world politics. The present day unipolar structure is therefore at best a transitory phase.

In the course of the short duration since the end of the Gulf War, it is amply clear that for all practical purposes we are living in a world of multinodal and multidimensional global politics, in which the three most important players (as of today) are: the United States, the European community, and the Commonwealth of Independent States (whose international posture counts, notwithstanding the loss of superior status). As Cohen (1991) analyzed it, all indications are that the Heartland state and China shall (for mutally compelling politico-geographic reasons) sort out their bilateral problems and build an international partnership of mutual cooperation vis-a-vis the other major power blocs. For the foreseeable future, China's global destiny appears to remain tied to the Eurasian Continental realm, which she and Russia jointly dominate. Despite recent strides in the country's economy, China's share in world commerce is only around 1.5 per cent, so that she can be no match to the rising economic clout of Japan in the affairs of Offshore Asia (to which the non-Communist littoral States of mainland Asia have in recent years, been increasingly drawn); Notwithstanding the decision of the Japanese parliament to restart sending armed contingents to other parts of the world to help maintaining world peace, international aspirations of Japan in the foreseeable future are likely to remain largely nonpolitical and nonmilitary in nature. Accordingly, rather than carving out an independent power bloc of its own, Japan may work toward building of a mutally beneficial alliance between Offshore Asia and the United States.

With the end of the cold war confrontation between the super powers, the post-1945 American foreign policy of "containment" has lost its relevance. Such a development has greatly reduced the strategic importance of South Asia for the United States. American military build-up in Pakistan is no longer in the former's interest. South Asia is, therefore, likely to be left alone to work out its own regional destiny through mutual cooperation and coexistence. In any such development of a regional formation in this area, by virtue of her commanding location, her size, and her relative economic strength, India is destined to play the pivotal role. By virtue of her leadership in South Asia, India is also assured of an important position in the comity of nations, particularly since, in view of the continued urgency to maintain an international forum through which to highlight the globally inequitous distribution of wealth among nations, and the fact that much of this wealth had been derived through crimes against the native people in the colonies, and from failure on the part of the colonial masters to pay due attention to the development needs of the colonies. Third world countries also need to jointly press for more favourable terms of international trade. History and geography have destined India to play the leading part in any Third World forum, just as it had been in the case of the Nonaligned Movement.

As regards Central and South America, these regions shall, in all probability, continue to remain under the domineering influence of the United States. With the last vestiges of imperialism rapidly disappearing from the scene, Sub-Saharan Africa is expected to continue with its economic and cultural ties with the respective "mother countries" in the maritime

European realm. Simultaneously, as part of the deprived world, this region is expected to maintain close diplomatic relationship with the South Asian realm, in general, and India, in particular. Australia should, in all likelihood, continue as a close political ally of the maritime European realm—with Britain in particular—with which its past ties have been so close that time has failed to dull the strong emotional bond. Economically, however, the region has already shifted to the Offshore Asian-American realm.

The faction ridden Islamic regimes in the Middle East are bound, for long, to remain a region of ongoing instability and mutual conflict, owing not only to the Arab-Israel problem (which is far too complex historically, culturally, and politically to get completely resolved in the immediate future, though the recent agreement between Israel and the supreme leader of the Palestine Liberation Front (in September 1993) represents a definite step forward in that direction), but also the Shia-Sunni division and political-military rivalry for regional supremacy between Iran and Iraq, and the current intra-Arab world problems generated by Iraq's invasion over Kuwait, leading to the breaking up of the fragile unity of the Arab world. Added to this are the problems generated by the current air of expectancy about the inevitable resurgence of republican revolutions in Saudi Arabia, the Emirates, and the rest of the dynastically ruled Arab States.

The Middle East is strategically important for all the three major geopolitical realms—North America-Offshore Asia, Maritime Europe, and Eurasian Continental—not only for its oil (which is certainly of great importance) but also as a valuable market outlet for manufactured goods and military weapons. The region is of added strategic significance to the Commonwealth of Independent States since it shares a long land boundary with Turkey and Iran, running over 2,250 km. Presence of sizeable Muslim communities, on both sides of this boundary, make it particularly sensitive. The Middle East, therefore, is the most volatile Shatterbelt in the present-day world, and it is likely to remain so for a long time. Overall, the situation is full of ominous possibilities, which may contribute to make this region a bone of contention between the United

States and the European Community. Such a development may adversely affect their functioning together as a single geopolitical bloc.

As Cohen (1991) noted, in the post-Gulf War world, Europe is likely to exert much greater influence in the affairs of the Middle East. Its influence over Iran, Iraq, and Turkey may become greater than that of the United States: "The lead taken by Britain and the European Community in proposing a 'safe haven' for Kurds in Northern Iraq is a case in point. So was the German effort to provide food and supplies to the Kurds in Iran. In both instances, the U.S. was pressurized to respond by supporting and adopting these initiatives. The region shall remain a Shatterbelt if the U.S. and the European community fail to forge a common agenda. Such dissensions might encourage the [commonwealth of Independent states] to re-enter the arena more vigorously" (Cohen, 1991). The potential in the Middle East situation to draw a wedge in the United States-European Community relationship are real and imminent. For any regime in the United States, it will be difficult to swallow a supporting (rather than leading) role in this important sector of current world politics. Any such turn of events may contribute to drawing together of the Commonwealth of Independent States (CIS) and the United States into a mutually rewarding partnership. While CIS needs American support to ensure for itself an active role in the affairs of the Middle East, for the United States, the CIS partnership may contribute to counter-balance the weight exerted by the European Community.

However, currently the most important sector of world politics is the bloc of States in Eastern Europe, which regained sovereignty following the dissolution of the Warsaw Pact. This region had been the traditional "Crush Zone" between the Germanic realm and the Russian Empire. The present- day States of Eastern Europe had originally been created to serve as buffer zones between Germany and the Soviet Union, in terms of the Versailles Peace Treaty (1919) at the conclusion of the First World War. In the years after the Second World War, they were incorporated into the sphere of the Soviet Union, as associated States, in terms of the Warsaw Pact. Their renewed status as independent States has, once again, transformed this region into a Shatterbelt, at least for the time being, even though in the foreseeable future there is no possibility of any armed conflict. Owing to the inherent nature of modern world politics to keep moving around the European culture realm in the Old World—to which the United States, Canada, Australia and New Zealand feel, for historical reasons, deeply attached at the-emotional level—the East European region is bound to remain a major focus of international politics.

There are three clear possibilities in East Europe. This group of States may develop into a loose federal union, towards which their history of common sufferings and rejoicings, their common administrative heritage during half a century of Soviet tutelage, and their urgent need for economic reconstruction and development may contribute to draw them. Such a development may also suit the geopolitical interests of the United States, so that America may be expected to promote such an arrangement. The European Community as well as the Commonwealth of Independent States, each may try to lure them into its fold. Germany should be particularly keen about it Apart from the weight of history, there are strong economic attractions for Germany in this industrially under- developed region lying next door. The region would offer a most valuable market for German goods. On both these counts, Germany should be keen on admitting them into the European Community. But any such move may prove to be a source of strain in the relationships between the member States. Britain and France may, in particular, be averse to any such idea. However, given the overall dominance of Germany in the affairs of the Community, any such German proposal may ultimately be carried through, especially since none of the other members are likely to be overly concerned about such a development.

Should this happen, the addition of so many new members sympathetic to the German line of thinking, may eventually make the Community an uncomfortable platform, both for Britain as well as France, since their position and voice in the Community might, in course of time, get marginalized.

Admittance of these States into the Community, therefore, may lay the foundation of eventual parting of ways between Britain-and-France, and the rest of the Community. In the case of any such eventuality, these two former colonial powers may work together to carve out a separate geopolitical bloc of their own. This may be achieved by drawing together the former members of their respective empires in non-Arab Africa. Given the complementary nature of the economies of the two mother countries, and their former colonies, there is every reason that any such geopolitical formation should succeed.

With the passage of time, memories of the colonial past have progressively become more of a binding rather than a distancing force between the former colonial powers and their post-1945 successor States. Any such British-and-French formation may inevitably become part of the global maritime realm led by of the United States. This shall be the surest recipe for the new formation to ensure that its voice should continue to count in the affairs of the Old World, and that the Middle East should never become out of bounds for the two 'European powers. Any alliance between Britain, France and the United States would" complete the formation of something like the Midland Basin envisaged by Mackinder in 1943. This should lead to a three-fold geopolitical division of the world into the Midland Basin, the Continental Eurasian realm, and the German realm of East and Central Europe.

There is a clear possibility, therefore, that the world may in the not too distant future, revert to a geopolitical scenario resembling the one postulated by Mackinder in 1943. The confrontationist concept of the Heartland (with its implied theory of perpetual conflict between the Heartland and the encircling Maritime Crescent) is dead, but the Heartland region appears nevertheless destined to occupy a permanent place in global politics. Mackinder's concept of "The Geographical Pivot of History" (1904) had drawn attention to a fundamental relationship in the political history of the Old World. Since Mackinder's concept of the joint landmass of Eurasia and Africa as the World Island has, in some ways, captured the essence of the post-Columbian reality of our world, the strategic significance of the pivot region, is destined to remain a permanent factor in global political and economic relationships, even though this region is neither the pivot, nor the heartland, in the manner Mackinder had originally envisaged.

GEOSTRATEGY AND THE WIDER CONTEXT : GEOGRAPHY OF INTERNATIONAL RELATIONS

Four Perspectives on International Relations

Wallensteen (1981) has identified four broadly non-exclusive perspectives in the study of international relations. First, "Geopolitik" which emphasizes the dynamic spatial setting as an environmental determinant of relations between States. Secondly, "Realpolitik" (or power politics) which concentrates attention on questions of military or political victories, the inventory of determinants of power being its chief concern. Under this perspective, all States, regardless of ideology, are assumed to act according to their hierarchical power ranging. Thus, most quantitative international research (QIR) is "realpolitik".

Thirdly, "Kapitalpolitik" (or economic relations) under which political and military relations between States are viewed as being guided by each State's economic self-interest. This is the basis of the currently popular world systems/model of Wallerstein (1979) under which researchers examine links between long cycles of economic growth and decline, the changing nature of resource competition and exploitation, and foreign policies of States. Lastly, "Idealpolitik", which examines the principles of legitimacy on which the State is founded and by which the State justifies its actions in domestic and foreign policy. For researchers in this stream, the twentieth century international politics has been like a game of chess between competing ideologies on the checkerboard of nations.

To Wallensteen, an overview of major events of world politics between 1861 to 1976, appeared to show that while the first three models explained 20-30 per cent of the conflicts, the "idealpolitik" model explained almost two-thirds of the total. Wallensteen made no attempt to explore the interconnection between the four-models.

It is quite typical of research investigations in international relations to engage in univariate explanations. This is the inevitable bane of the division of knowledge into "disciplines", so that each group of specialists confines itself to one specified circle of facts, and in that process the totality of the problem gets neglected. This was also the tragedy of our tradition in geopolitics and geostrategy. Both neglected complementary approaches to the study of international relations.

A distinguishing feature of contemporary research in international relations, however, is an increasing realization of the essentially complementary nature of the four broad perspectives summarized above. By imposing an orthogonal axis of an East-West scale (signifying division between capitalistic and socialistic States) on a North-South scale (signifying division between developed and underdeveloped, *i.e.*, or rich and poor States) we could visualize a quadripartite division of the globe into "Kapital" and "Ideal" politico-geographic spaces (Alker, 1981). Such a grouping of nations in two dimensional space provided us a good base both for geopolitical analyses through maps, or quantitative power-politics analysis through the simple addition of the strength of the rival blocs in the period before the end of the cold war.

Five Theories of International Conflict

In a 1984 review, O'Loughlin drew attention to five theories of explanation of international conflict. All five "have spatial analogues and when combined together into an interconnecting model, they could provide a useful beginning for a comprehensive approach to international conflicts". These theories are;

1. Bounding's (1962) *Theory of Viability* (also called "The sphere of Influence" approach) is a "realpolitik" model derived from Morgenthau (1973). It consists of identification and examination of zones in which a State may be dominated or is itself dominant. Foreign policy is viewed fundamentally as an effort to narrow down the extent of the former, and expand the latter. A major shortcoming of this approach lies in that it reduces complex international interactions to the simplistic notion that power is the key element in international politics, since aggression is basic to human nature.
2. Rumble's (1977) *Social field Theory* according to which nations are social units that interact in a social field. This interaction can be analytically divided into two sets of spaces—attribute and behaviour. In attribute

space, nations are located in terms of their characteristics (social, political, economic, military, and others) In behaviour space dyads of States are located in terms of their interactions (cooperation, conflict etc.). The two spaces are interrelated since the location of dyads in the behaviour space is a function of their location in attribute space. In a general way, the greater the sum of the distances in the attribute spaces, the greater the possibility of dyadic conflict. The emphasis on domestic attribute-international behaviour linkage in -this theory, somewhat parallels the call for a world-systems perspective in the study of political geography.

3. Sprouts' (1965) *ecological perspective*, which revived the environmental theme in the study of international politics. The two authors laid stress on domestic situation, foreign policy linkages, and pleaded for the inclusion of perceptions of the individual policy-makers as an additional element in the environment-international behaviour link.
4. Most forcefully advocated by Cline (1980), is the so-called *theory of power Politics*. It represents a further extension of the "sphere of influence" model of "realpolitik". Cline proposed a simple formulation for the study of foreign policy in which each State's power index is computed by employing the following formula:

$$(P_{p.} = (C + E + M) \times (5 + W))$$

where $P_{p.}$ is power-potential; C = Critical mass, (*i.e.*, the size of the nation-State), E = economic capability; At = military capability; S = national strategy; and W = national will. Cline's world view was a bipolar one, consisting of two competing blocs in the classic geopolitical fashion of the cold war era.

5. Cohen's (1973, 1982, 1991) *theory of Shatterbelts*, which maintains that certain regions on the globe, because of their complex ethnic, religious, political, ideological, economic, and physical composition, intrinsically stand out as areas of tension. They represent weak zones in international politics. Outside powers are attracted to

them by virtue of their strategic location and resources. As regions that do not clearly belong to one or the other power bloc, they tend to become centres of international conflict.

Toward a Time-space Model of International Conflict

O'Loughlin pleaded in favour of explicating the above five theories of international relations in relation to the space-time models in geography developed under the leadership of Hagerstrand (1967). The five theories can be viewed as associated with the study of five different sets of casual influences on international conflict, designated respectively as: distance, attributes, attitudes/beliefs hierarchy, and location. Each of these has been analyzed in spatial diffusion studies, and they have all been found to exert important influence on phenomena, both separately and collectively.

The influence of *distance* is modeled through distance-decay functions in diffusion studies. Such studies may be equated with the spheres of influence studies in international relations. In general, the greater the political/military/economic power of the State, the larger shall be the sphere over which it can extend its influence. *Attribute* (*i.e.*, domestic conditions) of the participating States in a given conflict, greatly influences their view of (and attitude toward) the conflict. This is the basic theme of much of the *social field theory* research. A spatial diffusion model can easily incorporate politico-geographic *perceptions* of the foreign, policy decision-makers.

These perceptions relate to the pattern of allocation of national resources, and the choice of technology to be adopted: "Just as certain farmers, by virtue of their attitudes, perceptions and beliefs, accept agricultural innovations while others reject them." Implicit in the power politics theory is the idea that States can be arranged in a hierarchy of classes, in accordance with their power status, such as: First Order (USA, CIS); Second Order (China, India, Japan); Third Order (Britain, France), and so on. Each level in this hierarchy corresponds to a particular geographical level of influence—global, continental, regional, and local. "Just as innovations move down the urban hierarchy as well as through spatial contagion, so, too, do international conflicts have both local (contagion) and extra local (regional or global) ramifications"—though international conflicts, unlike most diffusion process, may move up the hierarchy of power. (O' Loughlin, 1984). Finally, parallel with the corridors of movement in diffusion research, certain *locations* appear more prone to conflict because of the complexity of a host of historical, political, economic, ethnic-cultural, and positional factors. As O'Loughlin (1984) wrote: "Parallels between these theories of international conflict and spatial diffusion represent much more than coincidence.... They are representations of the same set of causal processes, a dynamism that lends itself ideally to space-time process models". (Discussed at length in Cliff and Ord, (1981)).

In a recent review of political geography O'Loughlin (1991) concluded that political geographers, like other social scientists, have to cope with global changes such as those that took place recently in the former Soviet Union, Eastern Europe and the Gulf. To tackle these, geographers need good process models that are contextually based. His message for the education of political geography student is that "Detailed analysis of contexts (in the guise of the new regional geography) must be blended with study of process models of change. In this way, perhaps, political geographers will have something useful to contribute to the understanding of future regional conflicts". This recommendation deserves a serious thought.

THE RIMLAND THEORY

Nicholas J. Spykman, born in 1883, was a Professor of International Relations at the Yale University and the director of the Yale Institute of International Studies. The most outstanding contribution of the American scholar was his book, 'The Geography of the Peace', published in 1944, shortly after his death in 1943.

In his book Spykman considered Mackinder's strategy of world domination contained in the heartland thesis and concluded that Mackinder's reasoning suffered from several weaknesses as he over-emphasized the potentialities of the heartland. Spykman discredited it as a possible centre of communication and power potential. Be pointed out that much of the area of the Soviet Union, lying to the west of the Yenisei river and forming the heartland was all wasteland.

He therefore asserted that the power potential of 'Eurasia lay not in the heartland but actually in the belt of states encircling the heartland on its west, south and east. This area Spykman named as the 'Rimland' which coincided approximately with Mackinder's 'inner or marginal crescent'. The American scholar considered the heartland less important than the rimland, which contained most of world's population and resources. He was convinced that the control of. Eurasia and ultimately of the whole world was possible through controlling the rimland. He, therefore, formulated his own theory, called the Rimland Theory :

"Who controls the Rimland rules Eurasia,

Who rules Eurasia controls the destinies of the world".

It will be seen that Spykman not only criticized the British geographer for his inaccurate heartland concept, but also suggested another strategy of world domination. But he has failed more miserably than Mackinder. No world power has so far been able to control all the countries of the rimland. In fact the task is not only difficult but impossible. Significantly, Spykman's rimland concept became the basis of the American "policy of contaminent" of the spread of communism in the coastal states of Asia.

MEINIG'S VIEWS ON HEARTLAND AND RIMLAND

Donald W. Meinig was critical of both Mackinder and Spykman and found fault with their over-simplified definitions of heartland and rimland. He wrote, "our definitions of heartland and rimland must......be rooted in cultural or I should like to term them....functional criteria". He noted that heartland was defined on the basis of interior location and hydrography. He, therefore, defined a new heartland which, though smaller in size than any of those delineated by Mackinder, was more stable yet functional.

More convincing and useful is perhaps Meinig's revaluation of the rimland, named differently as 'inner crescent' by Mackinder. Both the scholars considered whole of it as the natural realm of maritime powers, ignoring the fact that the orientation of the states, forming the rimland (inner crescent) had changed with time. It was 'outward' and 'maritime' during one period of history and 'inward' or 'continental' during some other period. Meinig termed the first as 'maritime rimland' and the second as 'continental rimland'. A state at one time close to the heartland power in terms of political ideology and trade might change its orientation under now circumstances. "One cannot (therefore) say in Mackinder's words that whoever controls the heartland rules the rimland— only that whoever controls the heartland is likely, under certain conditions, to dominate certain part of the rimland (inner crescent) and may aim at controlling it all" (Pounds).

3

GETTING TO THE SEAT OF POWER: THE GEOGRAPHY OF ELECTIONS

INTRODUCTION

Electoral geography is now more than 75 years old. The first published researches on the subject included those by Siegfried (1913) and Krehbeil (1916). These were followed by some other sporadic contributions (*e.g.*, Wright, 1932) and by the widely known Atlas of the *Historical Geography* of *the United. States* (Paullin, 1932), which contained a number of maps on voting results. Substantial geographical researches on elections were, however, made only in France, under the leadership of Andre Siegfried, who is rightly called the father of electoral geography. Siegfried's book (1949) on the geography of elections of the Ardeche region of France is a classic of its type.

Among the works of his numerous students and associates, perhaps the best known are those by Francois Goguel (1951) and Lancelot (1968). All these studies were based on the old man-land relationships tradition. With the rumblings of the quantitative- explanatory revolution in the social sciences, this traditional approach soon lost following. But, it was not immediately replaced, owing to the late arrival of the quantitative revolution in political geography (Dikshit, 1977, p. 234–239). Only in the 1960s, did electoral studies become an important theme in political geography. By the middle of the

1970s, many had begun to think that they comprised "the most activity developing aspect of political geography" and that electoral geography was now "one of the main sources of fresh insights into the general field of human geography" (Busteed, 1975). The popularity of elections as a theme for politico-geographical studies was evidenced by the publication of a popular monograph in 1975 (Busteed). This was followed by a detailed Penguin paperback on the geography of elections (Taylor and Johnston, 1979). Research on elections continued to be popular ever since.

Initially, the chief reason for this increased interest was that, in a general sense, more than any other aspect of political geography, the study of voting behaviour had begun to reflect the conceptual and methodological emphasis of the new mainstream of contemporary geographic research. This modern research was increasingly concerned with the use of sophisticated techniques of quantitative analysis, spatial diffusion processes, and behavioural studies. In 1968, a leading practitioner had gone so far as to declare:

> If political geography is to be integrated into geography as a science with an emphasis upon rigour in drawing hypotheses, hypothesis testing, and theory construction, then it is likely that studies of areal variations in the level of different political activities will play an important role in this integration. Data collected by bounded area) units referring to such area) phenomena as voting for different parties or issues are among the most available and accessible sources of quantitative data for the political geographer to apply to political geography canons of scientific methods used in other branches of the field (Cox, 1969b, p. 58-76).

PLACE OF ELECTORAL STUDIES IN POLITICAL GEOGRAPHY

In a 1976 review of Busteed (1975) Cox noted that a major underlying weakness of voting geography has been the failure to relate it to studies in political geography as a whole. The focus and objective of geographical studies in elections, -and the place of these studies in the total context of political geography, had not been adequately explained. In their geography of elections (1979) Taylor and Johnston explained how the contemporary definition of human geography as the study of spatial organization of society could be helpful in identifing the main geographical elements in a typical election. But they neglected to bring out the nature of the relationship between geography of elections and political geography as a whole.

The capitalist pattern of economy is essentially based on economic shortages, real or contrived. Owing to these shortages, the citizens experience a continuing conflict for the control of resources from which rewards (*i.e.*, satisfaction) can be derived. Because conflict is endemic in scarcity, capitalist political systems are essentially conflict resolving mechanisms (see Cox and Reynolds, 1974, p. 19–42). This theme of scarcity-and-conflict is central to the understanding of the problems of contemporary society. Much of the "new" geography of the 1970s is associated with what Smith (1977) called "the radical or relevance revolution" focused on such conflicts.

Since these conflicts are fought out in political arenas, much of contemporary human geography is partly political, since all studies related to spatial organization of society (or political organization of space) must inevitably be concerned with governments which are the chief determinants of "who gets what where"—the central question in all welfare-oriented geographical research. In all democratic countries with multi-party political systems, the electoral battle is the first step in the process of coming to power. Looked at from this perspective the geography of elections would appear to form the first stage in the study of the geography of politics in all modem democracies.

As Miliband (1969) argued, the capitalist class has to a great extent succeeded in organizing society and its control mechanism in its own interest. The political process of elections has, therefore, become of relatively minor importance—almost a facade—in the distribution of real power. Seen from this perspective, the main purpose that election serve is to ensure maintenance of the status quo, *i.e.*, to perpetuate the entrenched capitalist hold over the nation's affairs.

Ambitious individuals aspiring for political office may join a particular party because of their faith in its declared principles. But a good many people join the political party in which they are most likely to obtain power. Many would conveniently change parties if defections offer greater rewards. Such an unprincipled change of political loyalties, and the fact that successive elections achieve little more than the maintenance of *status quo*, may often breed indifference to electoral politics among the common people. This is all the more so, because candidates and parties tend to differ so little from one another.

Each party in a multi-party system has to be for ever conscious of the fact that in order to stay in power, it must accommodate the major initiatives of its rival party (or parties), so that it may be able to attract segments of non-committed voters, sympathetic to the opposite side. As Anthony Downs (1957) has explained, owing to this factor, political parties, like competing business houses, invariably converge in terms of their party proposals. That is why, changes in government in most democratic countries are rarely followed by major policy shifts. Party politics in most multi-party democracies in the non-socialistic countries has, in practice, "become little more than a contest for who shall sit on the horse already headed down a narrow lane with few bends and even fewer forks to be negotiated" (Johnston, 1977).

Despite these complications, politics continues to be of fundamental importance in modern times, since at present, in every political system, government controls increasing proportions of income and employment opportunities. Owing to this, decisions taken by governments regarding what is to be done and where, are of far-reaching importance in the life of the citizens. As such, holding the horse's rein brings immense power. And since the rein can be got hold of and firmly retained only through successful performance at the hustings, elections continue to be of great importance. While in federal States the process operates at three levels—centre, State, and local; in unitary States, it is only a two-level affair—central and local.

Viewed from this perspective, geographical study of elections is oriented towards the study and analysis of spatial aspects of how politicans at different levels of authority (local, central and state) obtain, manipulate, and maintain power. Electoral studies, therefore, are the key element in understanding the role of governments as spatial allocators. The nature of government—the spatial bases of support for the ruling party—is the chief determinant in the "where" part of the central question of "who gets what and where?".

If political science is the study of the processes of political decision making, then political geography could at least partly be defined as "the study of the spatial and environmental perceptions of such decision makers" (A.H. Taylor, 1973). It is with this aspect of political geography, that geography of elections is essentially concerned.

By introducing the valuable spatial component of "where" to the political scientist's central question of "who gets what", electoral studies in geography can contribute to a better understanding of the political processes involved in the process of getting to the seat of power.

The traditional view about the place of electoral studies in geography was that "uniformity in number of votes cast in favor [of] or in opposition to particular issues or candidates is indicative of spatial differences in voting. This lack of uniformity is of special interest to the geographer and should be as much an integral part of political geography as are the distribution aspects of other political phenomena" (Roberts and Rummage, 1965, p. 161-178).

But geographical study of elections is no longer limited to the depiction of areal distribution of votes. The political geographer today is equally interested in other spatial aspects of political behaviour manifested through the outcome of elections. As Rowley (1969, p. 398-400) wrote, a defining characteristic of recent developments within electoral geography as in geography as a whole, was not simply its empiricism but its concern for theory. Increased interest in electoral studies in 1970s was a logical product of the emerging trend in human geography, where the behaviour of the individual was being increasingly adopted as the basic unit of analysis, mainly as a reaction to the earlier concern with generalized data that had tended to disregard the underlying behavioural reality. Thus electoral geography became behavioristic in orientation.

APPROACHES TO THE STUDY OF THE GEOGRAPHY OF ELECTIONS

Approaches to the study of elections by geographers in the past may broadly be grouped into two categories: First, the traditional or the areal approach which attempted to analyze the man-land relationship involved in the spatial pattern of election results. This may in the current terminology, be called the "areal" approach. The second is the more recent approach emerging out of a dominantly spatial perspective (as distinguished from the areal) and the general behavioural impact on the study of political phenomena beginning around mid-1960s. This revised approach may be called "spatial-behavioural". The first is concerned with the "geography of voting" and the second with "geographical (*i.e.*, spatial) influences in voting".

Geography of Voting

Beginning with Siegfried's 1913 book, a dominant school of geographical study of elections had developed in France. So "vigorous was the work of geographers on French electoral behaviour that for long they have dominated this particular corner of French political studies, and their spatial insights and use of cartographic illustrations find a place in almost every study of elections in France" (Busteed, 1975). Siegfried's lead was followed, though much less vigorously, in other parts of Europe and the United States. The studies conducted prior to 1960s, belonged to, what Kasperson and Minghi (1969) called the "pattern analysis of 'behaviour in space' approach". Here, geographers concentrated on the temporal and spatial variations of voting aggregates and their association with environmental, social, and economic variables. Therefore, these studies were primarily concerned with the analysis of the aggregate election data by means of areal association in order to identify the significant environmental (including socio-economic) variables influencing the election results.

This traditional approach to the geography of elections was primarily concerned with the relationships between the total complex of physical and social environment of a region on the one hand, and voting patterns, on the other. It is best illustrated by Siegfried's 1949 study of voting in the department of Ardeche, on the west bank of the Rhone, during the period of the Third Republic (1871-1940). Siegfried analyzed and interpreted the pattern of party votes in so far as they mirrored the spatial organization of society which was itself believed to be greatly influenced by variations in the physical environment. The associations which he deduced were extremely constant. Those cantons that favoured parties of the left in 1871 still did so in 1936. In all, the 31 cantons were classified into 17 of the left and 11 of the right, with only three which varied greatly in their support during the long period.

The spatial polarization of political attitudes in Ardeche was, in a nutshell, explained as follows: The geology of an area is reflected by its relief (*i.e.*, altitudinal variations), and these two together influenced the type of economic pursuits followed in its various areal parts by determining patterns such as variations in farm size and population densities. These social and economic patterns influenced political attitudes, and hence also the election results (Fig. 3.1).

The essence of Siegfried's approach is, indeed, clear: "Through careful consideration of the appeals of the parties of the left and the right, he could identify those aspects of the social and economic milieux most likely to underlie the voting decisions, and was then able to trace those aspects to the physical environment" (Taylor and Johnston, 1979, p. 27). Siegfried did insist that there was a close "correspondence between the nature of the soil, agrarian landscape, type of dwelling, distribution of land ownership, degree of stratification in the society, the stronger persistence of tradition, and political orientation" (Dogan, 1967, p. 183). Nevertheless, it is wrong to think of him as a crude determinist.

He tried to explain electoral behaviour through correlations in the overall environment of a place, but he was far from an environmental determinist of the old style. This traditional areal approach to electoral geography consisted of two main streams. One group of studies concentrated on the spatial patterns of voting as depicted on maps, and the second, consisted of attempts to compare the voting map with maps of other phenomena—physical as well as socio-economic. The first group of studies may be called "areal-structural" and the second, "areal-ecological".

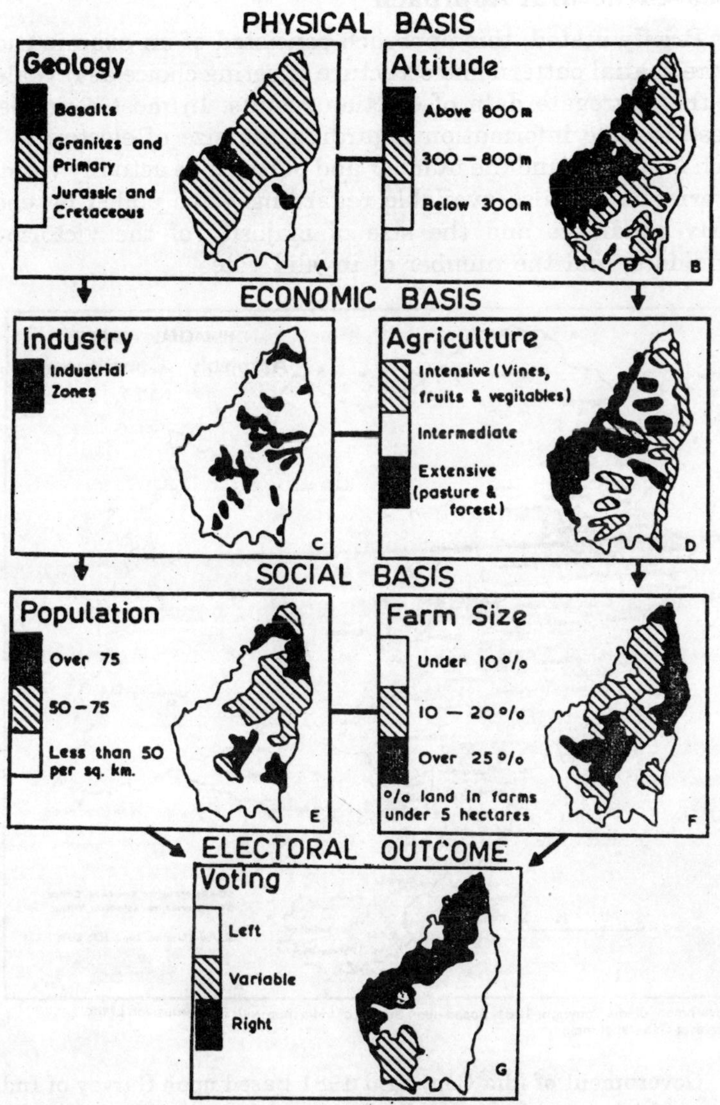

Fig. 3.1 : The geography of voting in Ardeche, France: Showing close relationship between physical, economic, and social bases of the region on one hand and the electoral behaviour of its population on the other (After Andre Siegfried, 1949.)

Areal-Structural Approach

Briefly stated, this approach consisted of an examination of the spatial pattern and structure of voting choice as revealed by the aggregate data of election results. In most countries, these include information regarding the size of electorate in each unit area, and the number and percentage actually voting. Information is also available regarding votes polled by each party candidate and the size of majority of the victorious candidate, and the number of invalid votes.

Government of Idia Copyright 1981. Based upon Survey of India map with the permission of the Surveyor General of India

Government of Idia Copyright 1981 Based upon Survey of India map with the permission of the Surveyor General of India

Fig. 3.2 : Assembly constituencies in Himachal Pradesh (India) 1977: An example extreme disparity in the areal size of electoral constituencies.

One way of using these data is to concentrate on spatial distribution of seats won by various contesting parties. This could be done either by means of a choropleth map or a map of symbols. Such a map gives a quick visual idea of the areal/ spatial pattern of party strongholds. But since this method takes into consideration only the cases of outright victories, it fails to measure the degree of support for the various parties in each case.

Whether it is the case of an overwhelming majority victory or just a marginal majority in a close contest where the victorious candidate might have won only by a small minority of votes, they are shown by the same choropleth shade or symbol. A constituency where the victory is by, say, over 50 per cent of the total votes registered, would normally represent the core of the party. The ones where the party candidates win only by small margins and poll well under half the total votes, may represent areas of marginal (peripheral) support. Therefore, in order to obtain a correct impression of the degree and durability of support for various parties, it becomes necessary that results of a series of successive elections should be mapped.

One of the major drawbacks of choropleth mapping is that the strength of minor parties or dissenting groups, is grossly under-represented. In order that a true picture of the division of support for the various parties in each constituency is depicted, it is necessary that the spatial patterns of voting in areas covered by each shade in a choropleth map are analyzed in detail. In most cases, in respect of parties with State or country-wide support base, the constituencies showing outright victories may represent only the peaks of a region of above-average support.

Another major difficulty lies in that the constituencies occupying large areas but supporting only small populations (usually in purely rural, underdeveloped, hilly and desert areas) loom large on the choropleth map and, thereby, convey a false picture of party dominance. For example, as Busteed noted, in the case of Scotland the rural constituencies of Inverness or Gallowy "achieved as much, if not more, visual impact as the thirteen small urban seats of Glasgow" (Busteed, 1975, p. 32). A most interesting case is presented by the example of Himachal

Pradesh where the area occupied by the mountainous constituencies of Lahaul-Spiti, Kinnaur, Bharmour, Baijnath, Kulu, and Banjar is almost equal to the area occupied by all the remaining constituencies of that state (Fig. 3.2). This difficulty can, to some extent, be overcome by resorting to the technique of map transformation suggested by Cotteret and Emeri (1957). Each constituency may be represented by a square of uniform size, and relative location of seats, and the overall shape of the area represented are as far as possible preserved.

This may often result in a highly odd shaped map, but the problem of variation in electorate-size and the area of particular seats is overcome. Further refinements can be introduced by varying the size of squares according to the size of the winner's majority or the percentage of votes polled by the victorious candidates. Alternatively, the base used for the choropleth map should be transformed on the basis of the size of voter population in each constituency (as is sometimes done to show the relative size of the countries of the world in terms of the size of their population (*e.g.*, Pounds, 1972, p. 129). This might help to rectify the wrong visual impression created by the usual choropleth map.

The importance of the areal-cartographic studies lies in the help they render in delineating regions based on political-ideological orientation. They help us in identifying areas of stress and strain in a given political system.

Areal-Ecological Approach

This branch of the traditional stream is related to the explanation of spatial patterns of election results in terms of their ecological context. The plan of these studies was simple: The scholars (including Siegfried)

> made the basic assumption that people will vote according to what they perceive to be their best interests. They then explored the characteristics of voters which might give clue to the nature of their self-interest. The area] variations in social and economic class, religion, nationally and race were among the prime factors considered. Using such methods, political geographers seemed to explain the foundations of voting patterns...

> [and] to predict how patterns would change as franchise altered or migration occurred (Prescott, 1972, p. 84).

The traditional approach to the problem consisted of using the cartographic technique of overlaying maps of election results over those of certain selected socio-economic variables considered by the researcher to exercise significant influence in the voting preferences of the electorate. Maps were often placed together (instead of overlaying) for comparison and visual areal correlation. Use of this approach may often yield very valuable insights and provide clue to useful hypotheses for further research. Siegfried's Ardeche study, cited earlier, was based essentially on this technique. Modem studies use statistical procedures, and do not trace the associations back to the physical environment.

A most distinguished contribution to the ecological-cartographic tradition was made by Lewis (1965) in his study of Flint (Michigan). He cartographically analyzed the relationship between income and Republican voting, and between Negro population and Democratic voting. This was an essentially a non-quantitative study and the methodology consisted of map comparison, but the maps drawn were highly sophisticated. The maps portrayed the distribution of Democratic votes by isopleths and the Negro territorial concentrations (from 1932 to 1962) by choropleths.

By using this technique, Lewis provided a visual exposition of the coterminous territorial expansion of the Black minority, and its impact upon the voting pattern. Although the voting data used were, in fact, areal data, Lewis could use them as point data for the purpose of drawing isopleths. This was made possible because the data were available for very small areal units which could be viewed almost as points on the map. This cartographic technique was subsequently used by O'Loughlin and Berg (1977) in their study of elections of Black mayors in the United States.

The method developed by Lewis has justly received praise from many students of electoral geography. The value of the technique is, however, rather limited, particularly in the context of countries where election results may not be published boothwise. Lewis's method assumes availability of socio-economic and electoral data on the level of very small areal units, rather like points on the map. A constituency is too large a unit to be conceived as a point. Another major difficulty is that Lewis had devised his method for a situation under which the population whose voting preferences were to be analyzed, had possessed marked areal concentration. Such a situation is difficult to come about when entire constituencies are to be treated as single units. Moreover, this method is applicable to a situation in which one presupposes high voter-turnout by particular ethnic groups, so that a single cultural-ethnic variable is able to explain maximum of vote variance. Then, as McPhail (1971 b) noted, "the interpretation of the patterns is by subjective comparison of numerous maps, and any attempt to isolate the underlying structural components of voting, are possible only at a speculative and subjective level".

The more recent attempts at areal-ecological study of elections, increasingly use advanced statistical techniques in examining correlations suggested by map comparisons of the type discussed above. Examples of successful combination of the cartographic and statistical techniques in the explanation of ecological relationship of electoral patterns are offered, among others, by Busteed and Mason (1970), Brunn and Hoffman (1970). Birdsall used choropleth maps and simple multiple regression techniques in his 1969 study of the "Preliminary Analysis of the 1968 Wallace Vote in the Southeast".

A wide variety of techniques have been adopted in this stream, but the most frequently used are factor analysis, or its related principal components form, and estimating equations of multiple regression type. Roberts and Rummage (1965) presented one of the earliest examples of the application of multivariate statistical technique to the geography of elections. They used multiple regression with residuals in examining the relationship between Labour vote in Britain in 1951 and 11 socioeconomic variables selected from the 1951 Census. Birdsall (1969) also tested his hypotheses by regression models while Cox (1968 d), Brunn et al. (1969), and McPhail (1970) used factor models which "are associated with a strong emphasis in recent studies upon factorial ecology (vide Ress, 1969), where in the structural dimensions of social and economic characteristic of a population in a designated area are identified as orthogonal components or factors" (McPhail, 1971b).

The advantage that the use of these statistical techniques has offered

> lies in their ability to deal with a large number of variables and a large number of observations. This is particularly true of factor and principal components analysis, where a large array of data is reduced by picking out those groups of variables most closely interrelated...[The] most useful quality [of these techniques] is that the groups of variables (factor in factor analysis, 'components" in principal components analysis) are picked out in descending order of importance and their significance in each area or constituency is indicated by a value or "score". When the variables include electoral statistics, the factors or components which contain the electoral variables may be regarded as those particularly with a political viewpoint (Busteed, 1975, p. 35).

But there are some very serious drawbacks inherent in the use of multivariate analysis in this context. One of these is the fact that these studies must rely on data sources which may not enumerate all the politically significant variables. For example, the Census reports can tell us very little about political history and tradition that are always significant in influencing people's outlook on politics.

Ecological Fallacy

With the increased availability of Census data, computation facilities, and package programmes, the possible dimensions and technical sophistication of doctoral studies based on aggregate data has considerably increased. There are however, certain in-built disadvantages in aggregate study of this type. In the first place, these studies lay stress only on the description of the spatial structure of voting behaviour. They either ignore or assume much of the underlying voting response processes. Secondly, the highly generalized voting data for entire constituencies cannot give the true picture of spatial distribution of electoral behaviour. Generalizations inferred from such studies cannot, throw any light on the voting decision of the constituent groups or the individual electors. Thirdly, any attempt to relate aggregate electoral statistics to aggregate Census data always runs the risk of falling into the trap of "ecological fallacy". Ecological fallacy may be defined as the error of assuming that ecological correlations at the constituency level are also correlations at the level of the individuals or small groups.

A classic statement on the problem of ecological fallacy was made by Robinson (1950) using data on Negro percentages and illiteracy rates within the American population. He emphasized the difference between "total" correlation (computed by taking individuals as units) and "ecological correlation" (computed on the basis of some aggregate units). But what Robinson had in mind was only one type of ecological fallacy, that is, the one that involves attributing to individuals the characteristics of the aggregate unit. There are many other types of ecological fallacies, such as the fallacy of attributing individual characteristics to groups, or "cross-level fallacies" that relate to generalizing about relations from only a small sample of the total population (Alker, 1969). The choice of areal sample is crucial. Each unit is a different ecological aggregate, as such, any generalization is valid only to the extent that the sample chosen represents a type.

As an illustration of ecological fallacy (in the sense that correlations calculated at one level of aggregation do not necessarily apply to observations at another level), we may cite the example of Jan Sangh vote in erstwhile Mysore State (Karnataka) cited by Dasgupta and Morris-Jones (1975). When we take districts as units, it is found that there is a good positive correlation between Jan Sangh vote and the distribution of Muslim population, *i.e.*, districts with higher Jan Sangh vote are also districts with large Muslim population. But this does not mean that the Muslims voted for the Jan Sangh. Since the figures given are at the level of districts it is not possible to discern how the Muslims actually voted. As the two authors noted,

> This result is in fact, also consistent with the opposite hypothesis that in the areas with large Muslim

> population the Hindus were more sensitive to communal issues and voted in larger numbers for a Hindu chauvinist party such as [the] Jan Sangh.

Despite the various kinds of fallacies that may be involved, ecological correlation has proved a most important method of analysis. Ecological correlation may be a poor substitute for "total" correlation based on individuals, but it provides valuable information about particular groups and places. Political changes are interwined with the total process of change and transformation in the entire social system. For this reason, a study of aggregates can be highly useful. To take an example, the land tenure system and its implications for the fortunes of various parties in an area is important on its own, even though we are unable to say, how landless labourers, share croppers, tenants and landlords voted. Moreover, as Dasgupta and Morris-Jones noted, group characteristics or contextual effects

> have their implications on the political behaviour of individuals. A particular mix of land system, occupational patterns, and demographical characteristics is likely to have a certain "compositional effect" on individuals. For example [in India], in areas with larger land holdings, the votes for the Swatantra party are higher than average, which is verified by...correlation analysis. This result does not imply that only the large landowners vote for Swatantra; if that were the case the Swatantra Party could hardly win any seat; on the contrary, it is more than likely that in areas with large land holdings, small tenants, landless labourers, and other groups, also vote for Swatantra. But the particular pattern of land holding prevalent in this area also typified a certain socio-political system under which individual voters operate. It is quite possible that the larger landowners make the decision about voting while the poorer men under them simply abide by these decisions (Dasgupta and Marris-Jones, 1975, p. 351).

Furthermore, in the study of political behaviour, it is not always the individual voter who is the focus of investigation. In the Indian context, for example, the individual is seldom the decision-maker, although it is he who slips the paper down into the ballot box. There is a lot to be said in favour of adopting household, village, tribe, or some such aggregate of individuals as the decision-making unit. Besides, ecological correlations are often of great value in generating hypotheses for further investigation. The affectiveness of ecological correlation depends upon the degree to which de jure units, for which voter returns are reported, coincide with the units for which the socio-economic data used are collected. The problem of unit of analysis has, therefore, to be suitably resolved, before a meaningful ecological analysis can begin.

Geographical Influences in Voting

Since the late 1960s, the study of voting behaviour began to clearly reflect the conceptual and methodological emphasis of new geography as the science of pattern analysis in space. Studies by such geographers as Cox, Reynolds, Johnston represented a clear shift in emphasis from areal to spatial, and from stress on locational attributes to a focused concern for patterns of relative location and spatial interaction. They emphasized that the traditional approach to the geography of elections was essentially an a-spatial one in the sense that the investigators

> abstracted the areal units from the space in which they were embedded and attempted to explain voting behaviour characteristics strictly as a function of within-area) unit economic and social characteristics; hardly any consideration was devoted to the spatial relationships of the area) units one with another and the impact of such spatial relationship upon voting behaviour (Cox, 1969, p. 112).

Expressing their dissatisfaction with this state of affairs, Reynolds and Archer (1969, pp. 2-3) wrote:

> Except in so far as the election results are often plotted on maps, the researchers adopting this approach considers each unit to be a multicoloured ball within an urn; the probability of drawing any one is independent of all other balls in the urn. No consideration is given to the spatial relationship of the

areal units one with another and to the impact of such relationship upon voting behaviour.

Like the more sophisticated statistical-ecological studies of the recent past, researchers following this approach gave full recognition to the multivariate nature of voting response. There was much less concern with the areal attributes of electoral constituencies and far greater emphasis on the measurement and analysis of such key spatial variables as distance, directional orientation, connectivity, and neighbourhood effect. In such studies individual voting behaviour is viewed as the result of the individual's particular characteristics as well as his inclusion within a distinct geographic milieu. The student attempts to explore how far factors such as distance-biased interpersonal relationships, residential relocation, and barriers to spatial interaction, have helped in shaping the voting response surface.

Cox had written extensively on the problem of identifying the mechanisms which determine voting surfaces. His primary concern was with the role of neighbourhood (*i.e.*, proximity) and spatial diffusion of political influences in voting. In a paper published in Progress in Geography, Cox (1969) attempted to relate voting decisions of individuals to their location in an information flow network. In behavioural terms, the individual voter may be conceptualized as a node or point in an information flow network. Explanation of the vote as a function of information flow, requires looking upon the node in terms of its sequential functions as;

(i) A sender,

(ii) An information processor transmitting and transforming received information, and

(iii) A receiver. Every elector is, in some ways, a sender of politically relevant information. But nodes are not alike in their propensity to send messages. Some individuals—the so-called opinion leaders—are more sought after as providers of advice than others. Besides, the senders may have significant partisan bias in the information transmitted, depending upon their political party affiliations.

This bias becomes significant in the role of the elector (node) as a transmitter and transformer of information, for it is this bias that would determine the nature of processing that the node performs. Receiving nodes require two types of information which jointly affect the development of a set of partisan attitudes. First, there is information which may be relevant for voting decision: information about political issues or about voting intentions of others. Second, a voting response to information is conditional upon receiving cues from politically relevant nodes—political parties or pressure groups. Without cues there is unlikely to be a voting response (Cox, 1969, p. 89).

The nature of information (*i.e.*, partisan bias) in a network depends upon the linkage associated with the flow of politically relevant information. Cox identified four bias-type relationships.

(i) *Geographical distance* bias, which describes a situation in which the probability of linkage between two nodes is a function of intervening distance;

(ii) A*cquaintance circle* bias, which States that "the probability of information passing between two individuals is related to the likelihood that they belong to the same acquaintance circle, be it formal or informal in character". Lower levels- of connectedness promote heterogeneity in voting decision; thus if the total network is divisible into relatively more connected subnetworks, the variance in the voting decision is likely to be reduced relative to the variance across them;

(iii) There is *forced field bias,* which means that certain elements are "endowed with drawing power" so that connections to those elements are more probable (Rapoport, 1957, p. 264). For example, drawing power could be a result of the greater political power of certain nodes or even a result of their ability to provide greater rewards than, alternative nodes.

In the nineteenth century rural England, for example, the loyalty of the tenant farmers and agricultural workers focused upon their landowners, in exchange for economic benefits which the landowner could bestow. Some such bias has been apparent in the influence that the traditional ruling houses in India have exercised over their former rural/tribal population in several parts of the country;

(iv) There is *reciprocity bias*, which has been defined as "the case in which the probability of link from a sender to a receiver is a function of a link between a receiver ,and a sender". Husbands and wives, for example, show very strong agreement in their political opinions, as do children and parents. Spatially, such reciprocal relationships are likely to be short distanced.

Operational Models of the Spatial-Behavioural Approach

The spatial-behavioural stream in the geography of elections was primarily concerned with spatial contagion and diffusion of politically relevant influences on the pattern of voting. These may, therefore, be termed as group membership studies. As Busteed summarized it, in the operational model of the spatial-behavioural approach, the group is regarded as an information-flow sub-system. The members of the group are the "points" in the information-flow network, and the membership of the group is viewed as a linkage with the network. The voting behaviour of a given group is regarded as an expression of group partisan attitudes and as a measure of the nature and volume of politically relevant information circulating within the group.

The Neighbourhood Effect

In studying voting behaviour in a social or group context, we focus on several milieux which impinge strongly upon most individuals in society. These include their work place and home neighbourhoods. Geographers have generally focused attention on the latter. It is well known that in a great measure, the geographical pattern of voting is related to the spatial distribution of various classes or income groups in a given society. In other words, a major influence comes from the spatial segregation of different social/income/occupation groups—a feature far more marked in cities than in rural areas.

Within each constituency, the map of social areas is often quite complex and contains as much variety as there is between constituencies. Each of these social areas in particular constituencies presents separate social milieu with a dominant political ethos of its own, although none of them is entirely independent of the other social areas surrounding it. These smaller social areas may,

> be grouped together to form catchment areas for schools, churches, and various social organizations, within which are brought together people of different backgrounds. So each little part of the country has its own dominant social ethos and political orientation. But no local population, however small, is entirely homogeneous in all of its attitudes, in part because the segregation processes are far from complete, in part because people move from place to place or from social group to social group and carry their learned attitudes with them, and in part because there always are mavericks, people whose attitudes are not consonant with the majority of their type (Taylor and Johnston, 1979, p. 234).

A complete analysis of neighbourhood effect would demand difficult and intensive fieldwork, for without that one cannot test the hypothesis that party support has greater spatial concentration than could be expected if all voters were not in contact with their local social environment. Three different strategies have been developed for the study of neighbourhood effect. These are:

1. *The Repeated Survey of a Large Sample:* This procedure was developed by three sociologists, Lazarsfeld, Berelson, and Gaudet (1944) who set forth to investigate how people decided whom to vote for in the ensuing Presidential elections in the United States. They chose the Erie County (Ohio) for this study as it seemed to represent the typical pattern of voting in the country as a whole. Six hundred respondents were interviewed once a month, from May until after the elections in November. Thus, a large bank of data was collected on the their preferences for the Democrat (Roosevelt) and the Republican (Wilkie) candidates. The methodology developed was used again eight years later in the small town of Elmira (New York) by Berelson, Lazarsfeld, and McPhee (1954).
2. *Single Survey:* The Repeated survey technique gives us a large and reliable data-base to work from, but it is

very expensive and difficult to design. Secondly, it is a major problem to keep the respondents interested enough to cooperate over a period of several months. For these reasons, many workers interested in testing neighbourhood behaviour, have relied on data collected through a single survey.

3. Aggregate Patterns: These studies are the ones most commonly pursued. They are based on published socio-economic data in Census reports. The data can be put together and correlated with election results, to bring out the effect of spatial contagion. For reasons of inherent deficiency in the set of data used, such studies can at best provide only circumstantial evidence regarding neighbourhood effects.

   Studies by Roberts and Rummage (1965), and Cox (1968a) were among the earliest attempts in this direction. One of the variables used by Roberts and Rummage was "distance to the centre of the nearest coalfield". This made the "spatial" orientation of the authors' study explicit. A most important contribution on the subject (though outside the field of geography) is by Butler and Stoke (1969, 1974) on British elections. Among geographers, reference may be made to the works of Biel (1972), Reynolds (1974), Cox (1971) and Johnston (1976 a and 1976 b).

The aggregate pattern studies under the spatial-behavioural stream (like the ecological ones) have primarily been concerned with relating election results to socio-economic data using similar statistical techniques. The difference lies in the sense that they have been designed within the framework of the principle of spatial contagion.

> Statistics describing socio- economic groups and election results are not regarded as stating that each individual in a group voted in a particular way. Instead the approach is to regard the statistics as descriptive of the most significant reference groups for sources of information and cues, with the election returns revealing the partisan bias of the cues which are conveyed (Busteed, 1975, p. 53).

It is truly said that within the whole field of geography of elections, the impact of neighbourhood studies may be negligible, since

> it is unlikely that it influences election results in many cases (because its basic finding is that anticipated majorities are exaggerated); it may be crucial in some constituencies which are divided into several social areas of different complexion, however; But it does have implications elsewhere in the subject, with respect to the spread of new parties and ideas, for example, to where parties should campaign to their best advantage, and to the geography of representation (Taylor and Johnston 1979, p. 267-69).

The Friends-and-Neighbours Effect

The friends-and-neighbours effect describes the situation wherein a candidate receives higher than average support in areas where he is particularly well-known. Whenever this kind of effect is discerned, a distance-decay mechanism is operative so that support for the candidate declines as one moves away from his home base. Much of the inspiration for studies under this stream was derived from V.O. Key's classic study on *Southern Politics in State and Nation* (1949) which had led him to inferences like the following (in the case of Alabama):

> Candidates for State offices tend to poll overwhelming majorities in their home counties and to draw heavy support in adjacent counties. Such voting behaviour may be rationalized as a calculated promotion of local interest, it also points to the absence of stable, well-organized, State-wide factions of like- minded citizens formed to advocate measures of common concern....In well-developed two party situation localism is minimized...... The classic case is that of Duchess County, New York, the home of Franklin D. Roosevelt, a Democrat of some note. The County, traditionally Republican, stubbornly held its partisan attachments and repeatedly failed to return a majority for even its most distinguished son (Key, 1949, pp. 37-38).

Key's entire analysis was based on choropleth maps and tabulations. Recent works in geography as well as in political science have used more sophisticated techniques. Reynolds (1969a, 1969b, and 1974) used Coleman's (1964) model to test the friends and neighbours influence on the voting response surfaces of a minor candidate in the 1954 gubernatorial election in Georgia (USA). A definite "home town" basis of support for the candidate was observed. The declining support away from the candidate's home town could be explained as a function of "increased per capita communication distance".

Reynolds' study emphasized interpersonal interaction space, rather than simple distance-decay. It also highlighted spatial clusterings rather than the independent groupings of the traditional approach. It is more of a behavioural approach as its concern is with the inference of information flow within a group, rather than elucidation of general political patterns. The relative ease which it can be operationalized, and the scope of incorporating a wider range of cross-sectional variables, makes the "friends-and neighbours" model one of high potential in future. [The only study of the friends and neighbours influence in Indian elections, is a case study of rural voting in the Sonepat parliamentary constituency in Haryana (Dikshit and Singh, 1993.)]

NATURE OF BEHAVIOURALISM IN QUANTITATIVE ELECTORAL GEOGRAPHY

Behaviouralism, with its emphasis on empirical research by analyzing the attitudes and characteristics of individual voters, dominated political science in the 1950s. It relied on data derived from individual survey research, facilitated by sophisticated polling methods that provided inferences about entire population from small samples of voters. Behavioural research examined relationships between voting behaviour and voter characteristics such as economic status, income, education, ethnicity, religion and occupation.

Behaviouralism in political science was closely-linked to a liberal, pluralist conception of democratic government. In brief, pluralism in democracy implied that governments are elected following open competition among organized political parties which offer alternative platforms to rational, self-interested voters.

> Thus voters select the party whose proposals are most consistent with their own views. Individual voter utilities and hence actual votes can be predicted on the basis of income, ethnic status and other economic and social characteristics. This pluralist conception of democracy was assumed to characterize Europe, North America and other industrialized democracies. Moreover, in conjunction with contemporary development theory, pluralism implied that an important component of the "modernization" process was the eventual adoption of Western democratic values in developing nations (Johnston, Shelley, and Taylor, 1990, p. 2).

Political behaviouralism idea was adopted as the central organizing concept of quantitative electoral geography in the 1970s and 1980s. Its research procedures were standardized for student use in Taylor and Johnston's *Geography of Elections* (1979).

ELECTORAL ANALYSIS AND THE REGIONAL CONTEXT

Western Europe (including its outposts in Australia and New Zealand), North America, and India are the major areas of the world where electoral studies have been pursued with vigour. In the Western countries, systematic and regular opinion surveys are carried out and behavioural study based on survey data is the predominant pattern of study. In India, on the other hand ecological analysis continues to be the dominant mode of study. The reason for this contrast lies in the fact that in most countries in Europe and North America, public opinion surveys are a regular feature. The techniques of random sample surveys have been developed to a point of accuracy that often permits close prediction of election results. Since opinion surveys have been conducted now for several decades, time-series analysis is also possible. As such, in most Western democracies, ecological analysis remains a useful device,

> principally for historical election studies, for describing the properties of an entire party system, and for analysis of the relationship between voting and context variables, rather than individual variables (Brass, 1978, p. 92).

The situation is altogether different in the case of India. The complexities of Indian society render systematic and regular sample surveys of the entire population both expensive and difficult to carry out. For this reason, sampling errors are unavoidable and indeterminate. Besides, contextual factors influencing the voters' choice in India are enormously varied. There is considerable regional variation in factors, such as the degree of persistence in patron-client relationships, relative ease of communications and transport, and patterns of inter-caste relationships. Consequently,

> the field remains open to the devotees of ecological analysis not only for time-series analyses, but for answers to the basic questions concerning the conditions for different patterns of competitiveness and turnout and concerning the socio-economic hasps of party support (Brass, 1978, p. 92).

SPATIAL ORGANIZATION OF ELECTORAL AREAS AND THE GEOGRAPHY OF REPRESENTATION

Every electoral system must be organized around a specific set of areal units covering the entire territory of the given State. This basic territorial organization and the precise definition of election districts or constituencies is a crucial element both in the entire process of election—party organization, campaigning etc.—as well as its outcome. The manner in which boundaries are drawn may greatly influence results, for boundaries may be devised with a view to favour a particular party or pressure group. This spatial manipulation of votes through constituency delimitation is known as electoral abuse and is based on the underlying concept of electoral bias. Electoral bias is measured as the difference between the percentage of seats that a party wins and the percentage of votes polled in its favour. If a party wins, "say, 50 per cent of seats and 40 per cent votes, then it is a case of 10 per cent positive bias. If the seats won are 40 per cent and the votes polled in favour 50 per cent, it would be a case of negative bias. Such electoral abuse in translating votes into seats is possible because different locations of constituency boundaries may produce quite different patterns of victories for the contesting parties. This may occur even if the underlying pattern of voting by different sections of the population remains unchanged.

Electoral abuse through boundary manipulation and geographical discrimination takes two basic forms. The first is called malapportionment which involves a design for constituencies under which different constituencies may enclose highly varying numbers of voters, with the result that a party with a much smaller numerical support in the overall population may succeed in winning a highly disproportionate number of seats. Following Taylor and Johnston, malapportionment may be illustrated with the help of the hypothetical example of a small country divided into five constituencies to elect five representatives. The Eastern Region of the country (the main base for the East Party, EP) is divided into three constituencies of 100 voters each. The Western Region (the main base for the West Party, WP) has only two constituencies, each with 150 voters (Fig. 3.3). This apportionment of constituencies would mean that EP has three safe seats out of five, and hence a stable majority, even though the overall numerical support for the two parties is equal. Under this arrangement of constituency boundaries even a marked swing (say of 10 per cent) of voter support in favour of the West Party (WP) is unlikely to grant her majority since the existing apportionment provides for a very high margin of support for EP (Fig. 3.3b). In most Western countries (UK, USA, Australia), malapportionment has been intimately linked with the issue of rural over-representation. This means that rural constituencies send their representatives for much smaller number of people per constituency, than the urban areas do.

The second form of electoral abuse involves the practice of drawing constituency boundaries in such a way as to maximize the electoral impact of support for one party at the cost of its rival. This practice of drawing constituency boundaries to favour one party over another (without involving any numerical discrimination in the size of constituencies) is known as gerrymandering, after the name of Governor Elbridge Gerry of Massachusetts. He had, through a redistricting law in 1812, created a pattern of constituencies (reminiscent of the salamander, a lizard-like animal-believed to live in fire) that defined their boundaries in such a way that his own party—Republican—Democrat could bag 29 seats with only 50, 176 votes. Its rival, the Federalists, could get only 11 seats even though they had polled as many as 51,766 votes (*i.e.*, 1,604 votes more than the victorious party).

Gerrymandering may also be explained with reference to the hypothetical example of a small country with a total number of four constituencies, as shown in the accompanying diagram (Fig. 3.4).

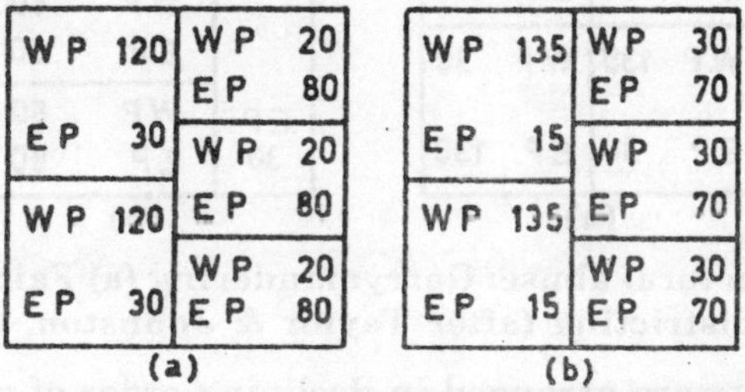

WP 120 | WP 20, EP 80
EP 30 | WP 20, EP 80
WP 120 | WP 20, EP 80
EP 30

(a)

WP 135 | WP 30, EP 70
EP 15 | WP 30, EP 70
WP 135 | WP 30, EP 70
EP 15

(b)

Fig. 3.3 : Electoral abuse: Malapportioment (after Taylor & Johnston, 1979).

In Fig. 3.4a depicts a situation wherein the support areas of the two rival parties (EP and WP) are so distributed so as to give a "fair" election result, wherein strength of party support is matched by the number of seats won. In Fig. 3.4b, the constituency boundaries have been manipulated (gerrymandered). Even though each constituency is of the same size (in terms of population), the net result of boundary manipulation is one in which the areas of WP support are divided in such a way that it gets an overwhelming majority in one constituency and a clear minority base in all others. Thus, gerrymandering by EP sympathisers has created a situation where three clear victories are assured to that party while its rival, WP, gets only one clear victory (albeit with a very large majority) even though the overall support for WP (60 × 3 + 130 = 310) is higher than that for EP (90 × 3 + 30 = 300).

Prescott (1972) suggested that there are two ways in which tests can be made to discover whether any gerrymander has occurred. First, we may inspect the map of constituencies to discover whether there are many boundaries with highly twisted shapes They may well have resulted from an attempt to include scattered areas of support for the favoured party. Secondly, the wastage of votes can be shown on a graph which records the percentage of votes gained by the parties for every seat contested.

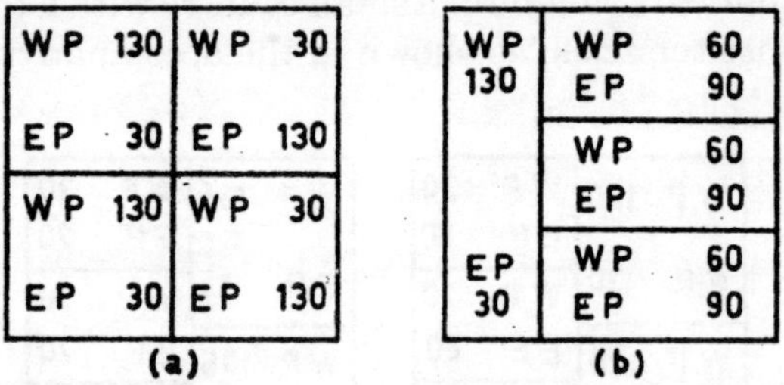

WP 130 EP 30	WP 30 EP 130
WP 130 EP 30	WP 30 EP 130

(a)

WP 130	WP 60 EP 90
	WP 60 EP 90
EP 30	WP 60 EP 90

(b)

Fig. 3.4 : Electoral abuse: Gerrymandering (a) Fair districting (b) Unfair districting (after Taylor & Johnston, 1979).

The seats are arranged in declining order of percentages. Figure 3.5 shows two graphs. The first graph depicts a situation where neither side is favoured; the second reveals that gerrymandering has favoured party B at the expense of party A. If through such a test it appears that one party has been favoured to the disadvantage of its rival, then detailed results of electoral subdivisions should be scrutinized in order to establish the nature and degree of electoral abuse. But, as Prescott cautioned, we must not jump at a rash conclusion, for it is perfectly possible that an apparent gerrymandering will occur even with honest application of sound principles.

It is difficult to evolve a standard for measuring gerrymandering. Some scholars (Bunge, 1966 and Schwartzberg, 1969) attempted to develop such a standard on the basis of the general argument that since:

> Malapportionment is reduced by restricting the range of size of constituencies from a standard population (the average quota)... therefore gerrymandering can be similarly reduced by restricting the range of compactness of constituencies from a standard shape (the circle) (Taylor and Johnston, 1979, p. 380).

This standard was most explicitly developed by Schwartzerg (1969) who suggested that the notion of compactness can be measured by the ratio of the perimeter of a constituency to a circle of equal area. The index for this standard has a minimum value of one, for a circle; the value increases-as the shape becomes more and more irregular. The whole concept of measuring gerrymandering on the basis of the compactness of shape of the constituencies involved is, however, based on wrong premises: Gerrymandering need not involve irregularly shaped constituencies:

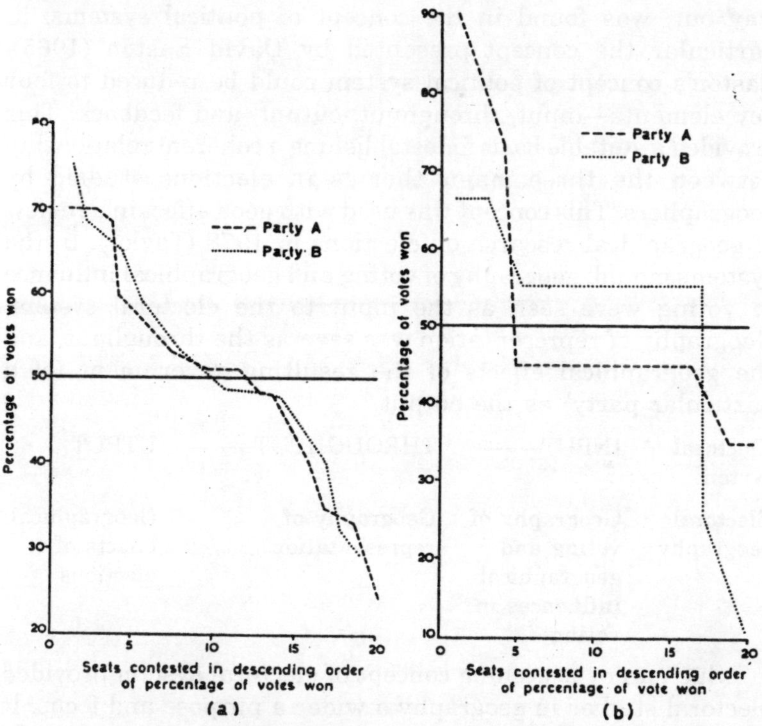

Fig. 3.5 : Measuring Gerrymander by graphical method.

> The shapes resulting from boundary discrimination will vary as the geography of votes varies. What circles, or any other abstract geometric shape, have to do with this relationship is problematic. Odd-shaped districts may facilitate gerrymandering, but this can only be assessed by considering effects on representation, not by measuring the district's degree of compactness (Taylor and Johnston, 1979, p. 384).

A SYSTEMS VIEW OFF ELECTORAL GEOGRAPHY

It is truly said that for all the research effort of the 1970s, electoral geography continued to be an uncoordinated and isolated discipline. The solution lay in finding a synthetic framework to integrate studies on the three major themes—geography of elections, geographical influence in voting, and geography of representation—into a coherent overstructure. A way out, was found in the concept of political systems, in particular the concept presented by David Easton (1965). Easton's concept of political system could be reduced to four key elements—input, throughput, output, and feedback. This provided a suitable basis for establishing a coherent relationship between the three major themes in elections studied by geographers. This concept was used with good effect in a review of geographical research on elections in 1978 (Taylor). In the systems mould, geography of voting and geographical influence in voting were seen as the input to the electoral system. Geography of representation was seen as the throughput, and the geographical effects of the resulting government (of a particular party) as the output.

Electoral system	INPUT——	THROUGHPUT——	UTPUT
Electoral geography	Geography of voting and geographical influences in voting	Geography of representation	Geographical effects of elections

Such an over-arching concept of electoral system provided electoral studies in geography a wider a purpose and focus. It also highlighted the fact that geographical studies in elections had remained largely restricted to the study of only one part—the input.

> By treating elections as an end in themselves, the actual purpose of election—the producing of legislatures and governments—had been largely forgotten. Hence the systems approach pointed towards a redirection of electoral geography beyond input and throughput (Taylor, 1989, p. 214).

Johnston (1980b) focused attention on electoral geography as part of political geography, viewed as the study as the study of the state as a political system. He saw the feedback loops between input and output as a step towards rounding off the system so that electoral processes represent an integral part of political processes. Johnston illustrated this in terms of the pork-barrel politics in America, wherein elected members ensure that the area they represent is well treated by the government, so that the electorate would repay them by their support at the next election.

In our discussion on the nature and functions of the modern State we referred to the two-fold functions of the State: one, as promoter of capital accumulation, and second, as promoter of social order through amicable resolution of inter-class conflict over the distribution of public goods. The two sets of functions of the State have become the basis of two related but different levels of politics. Gamble (1974) termed them respectively, as the politics of power and the politics of support.

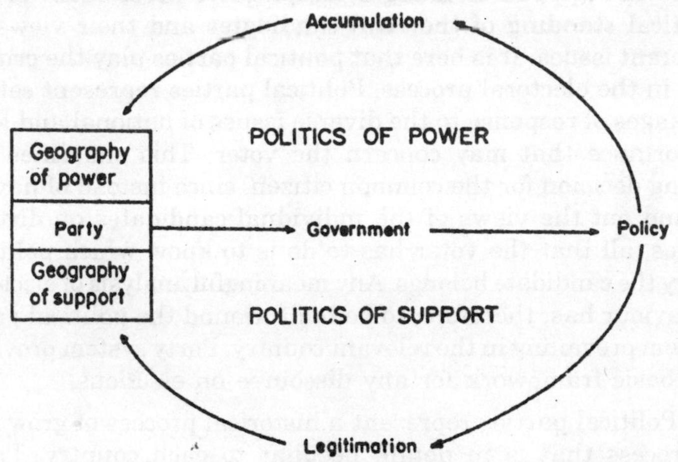

Fig. 3.6 : Electral Geography of Power and Support.

Since electoral battles are fought for gaining political control of the State, political parties are simultaneously engaged in both types of politics through electoral campaigns and political mobilization. Politics of power, relates to the promotion of special interests of the class base of the ruling party through suitable policy formulation. Politics of support relates to voter mobilisation through political and electoral campaigns. The operation of political parties in these two sets of politics give rise to two different types of electoral geographies—electoral geography of power, and electoral geography of support.

As Figure 3.6 shows, such a view of electoral geography replaces the single linear sequence in the systems model of electoral geography by two loops representing the two types of politics involved in every election (Taylor, 1989, pp. 226-227). Politics of power is essentially the politics of capital accumulation, whereas governments and ruling parties ensure their legitimation in the eyes of the electorate through the politics of support (Taylor, 1984).

THE SOCIAL CLEAVAGE THEORY OF ELECTORAL BEHAVIOUR

In any election, it is the individual voter who actually casts the vote in the exercise of his franchise. But the common citizen is too ill-ecqipped to make a comparative assessment of the political standing of the rival candidates and their views on different issues. It is here that political parties play the crucial role in the electoral process. Political parties represent sets of packages of response to the diverse issues of national and local importance that may concern the voter. This simplifies the voting decision for the common citizen, since instead of having to find out the views of the individual candidates on diverse issues, all that the voter has to do is to know which political party the candidate belongs. Any meaningful analysis of electoral behaviour has, therefore, to be built around the political party system prevailing in the relevant country. Party system provides the basic framework for any discourse on elections.

Political parties represent a historical process of growth—a process that is in detail, peculiar to each country. Party images—the "packages" that they represent—arc built over time and their development is intimately intertwined with the political events that have given character to the national present. A historical perspective is, therefore necessary in any comprehensive view of political parties. Modern liberal democracy as practiced the world over today, had its first beginnings in Europe, A brief introduction to the European party system as it has evolved over time is, therefore, necessary. The following discussion is based on the historical view of European party system presented by Lipset and Rokkan (1967), and Rokkan (1970).

Political party images and packages represent reflections of the social and political conflicts of the past. Conflicts produce cleavages that represent the opposite sides of the argument on particular issues. Conflicts and the associated cleavage patterns in any society can be studied in terms of two axes, representing respectively the territorial and the functional dimensions of conflict. The first produces conflict on regional lines, and the second produce conflicts that are based upon differences between rival all-nation interest groups on a whole range of socio-economic issues. Over time, these cleavages get translated into political parties that may represent opposing views on the issues involved. Such, for example, was the conflict between labour and capital in Europe in the post-Industrial Revolution period. This cleavage gave rise to new political formations under the banner of the various Labour parties espousing the cause of the working class. While the functional axis (capital versus labour) represents forces that lead to divisions across the nation—and may lead to the rise of national parties—the territorial axis (arising from place-specificities) explains the rise and survival of regional parties.

Political party systems in European democracies exhibit a good deal of variety. Lipset and Rokkan suggested four basic cleavages in European societies: the first two of these were the products of the National revolution, representing the process whereby the modern nation-State system had emerged. In the historical process of the rise of the modern nation-State, there were frequent conflicts between interest groups seeking, on the one side, to build a new centralized State, and on the other, local interest groups that were opposed either to the very idea of the State or to its basic principle, the *raison d'etre*, or both. The net outcome was conflict between the nation-building elite, on the one hand, and various ethnically, linguistically and religiously distinct subject populations (the peripheries) on the other. The nation-builders stood for a centralizing, standardizing, and mobilizing nation-State, in direct opposition to the historically established corporate privileges of the Church. Thus, the National revolution created two types of cleavages: one between the subject vs the dominant cultures, and the second, the Church vs the State. Political parties were organized to mobilize the enfranchised population on either side of the two sets of cleavages, and in course of time, they rose to occupy established positions in the structure and conduct of politics within each State.

A second group of cleavages was born as a result of the impact of Industrial Revolution, which represented the process whereby industrial capitalism became the dominant mode of production. The revolution gave rise to the conflict between the landed interests and the new class of industrial entrepreneurs. Among the landed interests, there was conflict between land owners and tenants; among the industrial classes there was conflict between factory owners and the workers. At the level of economic interests, there arose the cleavage between primary vs secondary economy (also referred to as town versus country) and secondly, between employers vs employees (generally referred to as class cleavage).

The four fundamental cleavages referred to above laid the foundation of the basic structure of political party systems in modern Europe. Since the impact of the two revolutions was felt at different times in different parts of Europe, the locally prevalent social and political conditions differed considerably from one country to another. Therefore, there is a great deal of difference in the precise character of parties in different countries in Europe.

Following the , extension of suffrage to the working classes, the workers vs employers cleavage became the most dominant one all over Europe. Prior to that, it is the other three that played the critical role in national politics. Extension of suffrage to the working classes gave rise to a series of Labour parties (or Left parties) all over the continent. Ever since, European party systems took a somewhat common complexion between the "left" and "right" parties.

The dominant governing elites all over Europe had, in general terms, alternative choices of alliances on religious or economic fronts, or a combination of both. The precise complexion of party system was the result of the socio-cultural structure of each country. Using the choices of National Church vs Catholic Church on the one hand, and urban vs landed interests on the other, Rokkan (1970), identified eight basic types of political party formations (or alliance-opposition structures). These were the outcome of three dichotomous divisions. The first occurred after the Reformation as a result of which there was either State control of national church (as in England) or the State was allied to the Roman Catholic church.

The second occurred after the "democratic revolution" (which ended the era of absolute monarchies). As a result, the first two types of party formations were further subdivided according to the strength of the established church in each country. The third came after the industrial revolution which created a division between those committed to landed interests vs those committed to urban interests. A further element in alliance-opposition structures was added as a result of the decisive thrust toward universal suffrage. Now to the cleavage structure in each country was added, the new element of class conflict.

Lipset and Rokkan concluded that the social cleavages analyzed in their scheme had proved both durable and stable, so that the cleavages found in West European societies in the 1960s were similar to those present 40 years earlier. And until recently, it had generally been believed that party systems in Europe had remained frozen at the level set in the 1920s. However, in a focused review of the Lipset-Rokkan model Johnston (1990, p. 121-141) has drawn attention to the increasing tendency toward dealignment in political party structures, resulting in the weakening of the traditional-loyalties between parties and voters. The recent electoral shifts in Great Britain are a clear indication of such dealignment.

The 1987 general elections in New Zealand also revealed political realignment based on restructuring of traditional political cleavages. The rising trend toward dealignment has cast doubts on the supposed durability of traditional cleavages since the 1920s, and points towards the rising influence of local and regional factors in the shaping of political attitudes and party identification. As Johnston noted, long term stability requires short term flexibility. In his examination of the recent shifts in voter alignments in Great Britain, Johnston found that while, in general, cleavages aid in structuring political conflict, those which are too rigid frustrate the possibility of minority victory. Such a frustration tends to create long-run instability in the national political system. The restructuring of traditional political cleavages-is a response to this contingency.

PROGRESS IN ELECTORAL GEOGRAPHY

Notwithstanding its great popularity as a field of inquiry throughput 1970s and 1980s, electoral geography Tailed to make news. Although the quantity of research production was impressive, quality-wise, there was not much ground for jubilation. Theoretical insights were few and far between, so that geographical contribution to electoral studies in general have failed to generate appreciable enthusiasm among other social scientists. The aim of most studies made by geographers was nothing more than an understanding of the electoral result in the context of the local/regional factors operating at that time.

> The result has been a general failure to link geographies of elections together into a coherent body of knowledge... On the whole quality has lagged behind quantity of production and few would now claim electoral geography as a success story (Taylor, 1989, p. 203).

Since there were hardly any attempts at comparative study of elections in different countries, the geographical explorations into electoral analysis remained largely isolated and uncoordinated. Owing to this, "the geography of elections is a topic sorely missing from quantitative electoral geography". Taylor's "world systems interpretation of elections" was an attempt to redress this research gap.

THE WORLD-SYSTEMS THEORY OF ELECTIONS AND THE CRUCIAL CASE OF LIBERAL DEMOCRACY IN INDIA

The purpose of this section is to examine the validity of Taylor's (1986, 1989, 1990) world-systems theory of elections with special reference to India, which Taylor described as the crucial test for his theory. This discussion is divided into three parts. Part one presents the main thrust of the theory and its author's perception of the democratic process in India. Part two is devoted to a critical review of Taylor's analysis of Congress voting patterns, and extends the analysis to the full range of nine parliamentary elections held between 1952 to 1989 in order that the nature of continuity and change in electoral patterns may be suitably highlighted. The final section summarizes the findings of the discussion, and attempts to explain the special nature of Indian democracy.

In 1986, Peter Taylor presented a world-systems analysis of political parties. This became the basis for his world-systems interpretation of elections (Taylor, 1989, chapter 6) which maintains that, rather than developing separately in each country, politics are part of the larger, world-wide, system. Most political parties represent a set of ideas which are inevitably linked to certain broader ideas and ideologies common to many countries and their political parties. Interpretation of an idea or doctrine may vary from one country to another, but political parties cannot be autonomous of the political world beyond their borders.

The theory maintains that free and fair elections are an essential ingredient of modern liberal democracies committed to social welfare. By its very nature, the liberal democratic process through periodic elections can succeed only in the relatively rich and developed economies as in Western Europe and North America. The rich economies of these industrialized countries have enabled them to develop a politics of redistribution. Such politics may not be possible in most countries in the underdeveloped world. When modern party systems took root, the States in Western Europe were rich enough to allow meaningful competition between political parties on the distribution of the "national cake", where all citizens could potentially benefit.

As such, elections became keenly contested political battles since the question of "who gets what" encompasses all sections of society. Owing to the depressed state of the national economies in the Third World, such an exercise is unlikely to succeed, since the scope for the politics of redistribution is highly restricted. There is little for a ruling political party to redistribute and thereby retain loyalties of their voters intact. This inevitably leads to an indifferent attitude toward electoral contests. As Taylor wrote, "why should the vast majority of a State's people participate in an election if there is little or no politics of redistribution?"

Since in the Third World countries, political parties elected to form government are hard pressed to pursue a politics of redistribution that can retain their voter's loyalties, geographical pattern of party support may keep changing from one election to the next. This is because parties are unable to maintain the support of those who have supported them in the past. Therefore "in order to study elections world-wide, we need to bear in mind the very different politics resulting from huge differences in material well-being between countries in the core and in the periphery" where a very different political process seems to be going on (Taylor, 1989, pp. 223-224).

Referring to India, Taylor observed that in his world-systems interpretation of elections,

> India cannot be a liberal democratic State because it does not have the resources for a viable politics of redistribution. But it does have regular free and open elections. If Blondel (1978, p. 55) is correct [in maintaining that, 'India constitutes the largest liberal democracy in the world'] this would be the exception to prove our world-systems analysis wrong. Hence Indian politics represents a crucial test for the world-systems theory (Taylor 1989, p. 246).

With detailed references to electoral geographies of the Indian National Congress, Taylor tried to demonstrate that the electoral politics in India is very different in nature and substance from what obtains in the West, so that his theory holds true in the case of India as well.

Taylor drew attention to the fact that one of the general features of the successful mass parties in the countries that became independent after World War II, is that after independence their mass support base seems to have disappeared. According to him,

> In the case of India, Congress found it had no stable mass base but was able to survive by transforming itself into an aggregative party. In order to create aggregations the party had come to rely on purposive ambiguity in its policy positions....[which] allows a broad range of elites and voters to be accommodated in the party fold in a Congress dominated party system (Taylor 1986, p. 13).

Taylor maintained that the Congress party had failed to develop a well-defined doctrine or ideology behind which it could build a national constituency of support. Therefore, the party could not perform the electoral-integrative role that parties in western liberal democracies do. This, in his view, implied, that at any particular election, the Congress party's support base is "a mere aggregation of different groups" whose composition may keep changing from one election to the next. The net result is that unlike the mass political parties in liberal democracies in the West, Congress lacks a "normal" vote pattern.

Taylor built his case for a World-systems interpretation of electoral politics in India on:

(i) A case study of Congress voting pattern in the assembly elections in Punjab, 1952-1972 (drawing on Dikshit and Sharma 1982, and Brass 1975), and

(ii) A comparative view of the all-India distribution patterns of "above average" and "below average" voting for the Congress party in the parliamentary elections of 1971 and 1977. His comparative analysis of voting patterns in the two elections led him to conclude that voting patterns in India show extreme geographical instability. "In 1971 the pattern of support is concentrated in the Hindu heartland of the north. In 1977 the north is lost and most Congress votes come from the non-Hindu [sic] southern States producing a completely new pattern of votes". This has led Taylor to assert that his theory applies to India, and as in the case of other Third World countries, in India also, "stable congruent politics is not a possibility".

In regard to electoral performance of the Congress party in Punjab, the author emphasized that the ability to create alternative coalitions is the hall-mark of an aggregative party. In Punjab, "whereas the Sikh party, Akali Dal, and the Hindu party, Jan Sangh have stable patterns of support, Congress alternates its support between the two religious communities". Owing to this Congress voting patterns keep oscillating since "as the national aggregative party Congress has been able to shift between the two religious communities for its particular electoral advantage. In electoral geography terms there is no 'normal' vote for Congress" (Taylor 1986, pp. 5-20).

The rest of this paper analyzes the veracity of Taylor's evidence.

A Critical Review of Taylor's Evidence

Taylor failed to appreciate that Assembly and parliamentary elections relate to two very different sets of political battle fields requiring very different campaign strategies. In the case of the former, all contesting parties must project themselves as promoters of regional interests and identities. Their policy programmes have to be adjusted to the specific regional context. This becomes necessary to counter the appeal of regional parties who have relatively little stake in all-India political contests. Furthermore, voter response in the two sets of elections are expected to differ. The electorate may prefer to vote in favour of a popular regional party in assembly elections in the hope that it shall be more responsive to local/regional problems. But the same electorate shall most often be reluctant to vote for such a party at the parliamentary elections. Since, owing to its restricted representation, a regional party may often prove to be ineffective in bargaining development funds and other concessions for the state from the centre. Electoral politics under federalism is a two-level involvement—the two levels, though interdependent, are largely autonomous. For this reason inferences drawn from electoral performance of political parties at the state level shall in most cases not apply to party performance at the national level (Dikshit and Sharma, 1993).

Secondly, Punjab is by no means the right choice for understanding the trends and patterns of state politics in India. It is the only state (besides Jammu and Kashmir) where the electorate is sharply divided between two conflicting communities, each having a sizeable presence, and each identified with rival political parties engaged in promoting the interests of their respective communities. In Punjab, the two parties are Akali Dal (Sikh) and Jana Sangh/BJP (Hindu). Inferences drawn from state politics in Punjab are, therefore, not likely to be repeated in any other major state unit.

Congress was by no means above caste and community considerations in electoral calculations, and it had all through deliberately maintained a vote bank among the Muslims and the Scheduled Castes and Tribes. But being a centrist party with high stakes in national politics, Congress could not afford to be seen as a party engaged in creating disenchantment between rival communities by favouring one community against another.

The spatial pattern of Congress vote in Punjab oscillated from one election to the next not because it lets down each community that votes for it in turn. Far from it. The Indian National Congress originally represented the movement for freedom. After it converted itself into a political party following independence, it inherited the untrammeled image of secularism (with a sympathetic tilt toward the minorities) from the parent organization which had won the country freedom from colonial rule. In the course of the struggle for freedom, Congress had brought diverse communities and pressure groups together under its organizational umbrella to work unitedly for the common goal. It is in this sense, that Congress has been viewed as an aggregative party—one that successfully organized diverse ethnic-cultural communities, and conflicting economic interests, to work together under its organizational umbrella for the achievement of freedom from foreign rule. This distinctive aggregative style was retained after independence since the party had no rival in the national political arena to contend with.

So long as Congress had symbolized the struggle for freedom from foreign rule, there could be no scope for any alternative secular party to appear on the scene. Freedom had come virtually unannounced, and all too soon, following the installation of the Labour party government in Britain, in 1946, to allow time for rival political parses to organize and build durable support bases of their own. The old aggregative character of the Congress party therefore continued long after independence had been achieved. Under the Congress dominated system (that lasted until late 1960s) political panics in the opposition had very limited role to play. The opposition parties fought elections not in the hope of forming a government at the centre, but in order to be able to send to the parliament their own representatives who could influence sympathetic sections in the Congress party's policy formulation team, to promote their specific group interests. (On Congress-system see Kothari 1964, and Morris-Jones 1964, 1966).

The very nature of the Congress party and the country's socio-political environment made it inevitable, that the party's policy programmes would not be ideological but Instead aim at all-round development of the State and its constituent regions and communities. Apart from the Communist parties (whose effective support base is confined to Kerala, West Bengal, and Tripura) and the rapidly expanding Bharatiya Janata Party (the new incarnation of the old Jana Sangh) India had never had any significant all-India party with a clear cut ideology. Both the Janata Party (1977) and the Janata Dal (1989) have been (in ideological terms) poor imitations of the Congress party, to which most of their more prominent leaders had once belonged.

The Congress party (like political parties in liberal democracies any where else) is the product of the specific history of political development in the country since 1919. In that year the party had launched upon the course of mass mobilization in the cause of freedom from colonial yoke. And as in the case of political parties in Britain, in the case of Congress also, "The actual organization operating at any particular election.... is not normally designed for that election" only, and "The manipulation of the political agenda is not a conspiracy of ruling elites but rather reflects the relative power of different interests in the evolution" of the party. At least, this was true of the Congress party until the split of 1969. Upto that time, elections to party posts from grassroots to the highest level were regularly held. Not- withstanding Jawaharlal Nehru's towering presence, the party was nobody's personal fief. Things did change after Mrs. Indira Gandhi took over the command—

but the change was restricted to matters pertaining to party organization. The overall perspective and policy programme of the Nehru era had been retained.

Taylor's article carried a graphical display of correlation between Congress voting and per cent Hindu population in Punjab. The graph showed positive correlation of over 0.4 in 1952, 1962, and 1972. A weak correlation of 0.2 had appeared in 1969, but the picture had been completely reversed in 1957 and 1967. In these two elections, Congress voting was negatively related to per cent Hindu population, the correlation values being -0.6 in 1957, and –0.4 in 1967. The reason for this inconsistent relationship was that: In 1952 the voting pattern in Punjab was in line with the general trend across the country. Congress received support from all classes and communities. The huge negative correlation value of -0.6 in 1957 had been caused by the merger of the Akali Dal into the Congress party. The Hindu population of the state did not take this merger kindly and, therefore, voted against the Congress. This marriage of convenience was rather short-lived. The two parties parted company in 1960. This gladdened the heart of the pro-Hindu elements, and in the 1962 elections they cast their vote in favour of Congress. Correlation between Congress vote and per cent Hindu population therefore, returned back to the 1952 level.

The complexion of party politics in Punjab had changed drastically following the decision of the Congress government in Delhi to concede the Sikh demand for Punjabi Suba (the present Punjabispeaking State of Indian Punjab). Hindus in Punjab had all through opposed the demand for division of the state on linguistic lines. They were angry with the Congress leadership for having sacrificed their interests, and voted against the party at the 1967 election. The intensity of Hindu resentment was reflected in their massive negative vote so that in 1967, correlation between per cent Hindu population and Congress vote sharply declined to -0.4. Hindu resentment against Congress continued for long, though with the passage of time, its intensity was reduced. As such, at the 1969 election, the correlation value became positive (0.2) though it continued to be weak. In the election of 1972, Congress was riding the crest of popularity following Mrs. Indira Gandhi's successful intervention in the struggle for liberation of Bangladesh leading to the dismemberment of Pakistan. All sections of society (Hindus as well as Sikhs) voted for Congress in large numbers. And the correlation value of the relationship between Congress vote and per cent Hindu population reached the highest level.

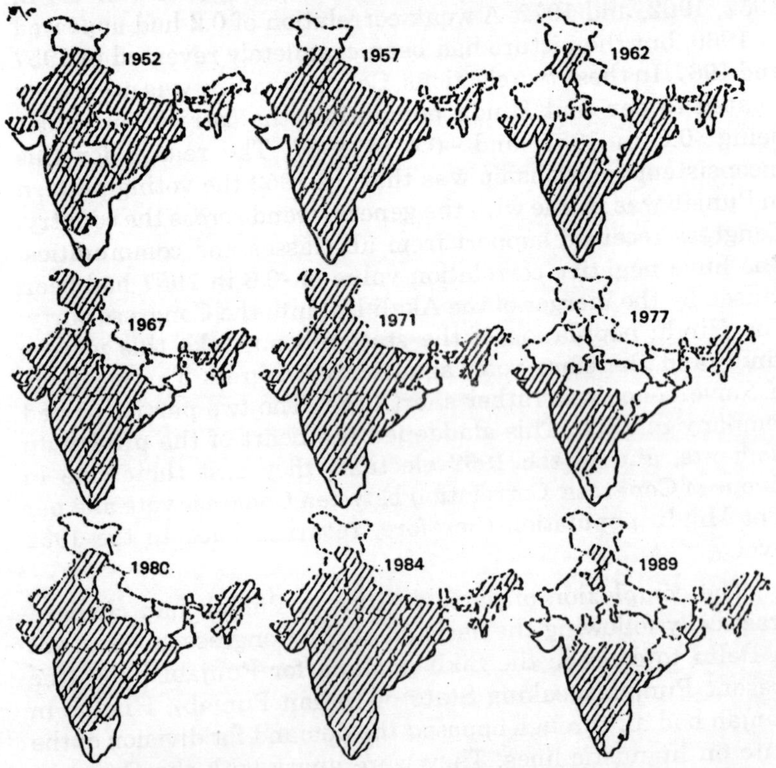

Fig. 3.7 : Parliamentary elections, 1952-1989: Areas giving above 40 per cent vote to Congress party (Data by constituencies).

Taylor compared and contrasted the electoral results of the 1971 and 1977 parliamentary contests with a view to reinforce his argument that the (i) the Congress party is essentially a limited purpose alliance whose complexion keeps changing from one election to the next, (ii) the party is in the habit of letting down each community that has voted for it, in turn, and (iii) the party keeps countering voter rejection at the next election by creating a new support base (with the result that spatial pattern of the party's vote keeps changing). Above average and below average electoral performance of the Congress party were separately plotted for the two elections (taking state as the basic unit). The maps showed that whereas "In 1971 the pattern of support is concentrated in the Hindu heartland of the north. In 1977 the north is lost and most Congress votes come from the non-Hindu[sic] southern States producing a completely new pattern of votes: these were turnover elections in terms of geography as well as politics" (Taylor 1989, p. 248).

Table 3.1 : Electionwise total number of runner-up candidates polling above 40 per cent votes.

Election Year	*Total number of runner -up candidates with above 40 per cent votes*	*Total number Parliamentary Seats*
1952	26	489
1957	106	494
1962	116	494
1967	116	520
1971	100	518
1977	149	542
1980	91	542
1984	168	542
1989	167	527*
Total	1039	5068

*In 1989, elections were not held in 27 constituencies falling in Assam (14), and Punjab (13).

It is not clear as to what the author means by "average" vote. If average here implies the percentage vote share of the Congress party for the whole country in a given election, then it is not a satisfactory measure to judge the overall standing of the party in the individual states. For example, in a particularly favourable election year (as in 1971 when Mrs. Gandhi was riding the crest of popularity wave, or 1984 when owing to her brutal assassination there was a strong wave of sympathy in favour of her son) the Congress party may poll unusually high percentages of vote in areas of generally unstable electoral politics (such as those in the Hindi belt). But states with more stable electoral politics may not be overly affected by such occasional events, and may continue to maintain their usual voting levels. During a pro-Congress popularity wave, therefore, the latter group of states may record lower than average Congress performance in that electron (even though their own pattern of voting itself may not have changed), leading to a false impression that the election was a turnover one.

Constituency level analyses of election results for the nine parliamentary elections held until 1989 has revealed that in elections to the parliament, candidates polling above 40 per cent of the total valid vote have a four-to-one chance of victory (Fig. 3.7). When the results for the nine elections are clubbed together, it is found that on an average, in only about 20 per cent of the consitutencies, candidates polling more than 40 per cent votes had failed to win. (Table. 3.1) A vote share of forty per cent may therefore, be taken as the winning margin for elections to the Indian parliament. The accompanying map, depicting states with above and below 40 per cent voting in favour of the Congress party, reveals that except for the unusual election of 1977—when the very survival of democracy had become the electoral issue—and in that of 1989—parliamentary elections in India have not been "turnover" ones.

The map shows that until 1967, Congress dominance of the electoral scene was nearly complete. Only Madras and Madhya Bharat in 1952, only Kerala in 1957, Kerala, Rajasthan, and Uttar Pradesh in 1962 had recorded less than 40 per cent vote in the party's favour. The passing away of Nehru in 1964 had adversely affected the strength of the Congress party. In 1967, the first election to be held after Nehru's death, the number of states recording less than 40 per cent voting in favour of the Congress had substantially increased. The list now included Bihar, Orissa, Punjab, and West Bengal, as well as Kerala and Uttar Pradesh (U.P.) both of which had figured in the 1962 list also. Three of these states—Kerala, Orissa, and West Bengal—again gave less than 40 per cent vote to Congress in 1971. The state of Tamil Nadu was the only other State to join the list. The 1977 election was of course a real turnover one—all states except Gujarat, Maharashtra, Karanataka, and Andhra Pradesh—had recorded less than 40 per cent vote for the Congress party.

The collapse of the Janata Party within two years of its birth, and Mrs. Gandhi's public expression of sincere regrets for having imposed the Emergency, and her promise not to repeat it, re-established the party's old support base. In 1980, Congress polled above 40 per cent votes in every state except Kerala, Bihar, Haryana, J & K., U.P., Tamil Nadu and West Bengal. In 1984, except for the three problem states for Congress—Assam, J & K., and Punjab—only Kerala had given the party less than 40 per cent vote. In 1989, in the whole of north India Congress votes declined to less than 40 per cent but the usual voting pattern had continued in Maharashtra as well as the four states in the south.

Scrutiny of the list of states voting below 40 per cent in favour of Congress since 1967 (when electoral politics in India had entered a new phase, marking the end of the era of absolute domination of electoral politics by the Congress party), we find that in course of the six elections (held between 1967 and 1989) the states of Gujarat (except in 1989), Maharashtra, Karnataka, and Andhra Pradesh have consistently given above 40 per cent vote share to the party.

As regards states giving less than 40 per cent vote share to the Congress, the position was as follows: Kerala six times out of six, West Bengal 5 times out of six, Bihar, U.P., Punjab, Orissa, and J.& K., each four times out of six, had given less than 40 per cent vote share to the party. This would hardly justify calling these elections as turnover ones. Turnover had occurred only twice—in 1977 and 1989—each time following the formation of a strong anti-Congress front by the opposition parties. It may also be noted that the states of Gujarat (except in 1989), Maharashtra, the four States in the south, and West Bengal have been all along consistent in their overall voting pattern. Is it that the world-system thesis on elections applies only to the northern half of India and particularly to the Hindi-speaking states?

Conclusion

To conclude, a close scrutiny of state elections in Punjab as well as all-India elections to the nation's parliament, shows that Taylor's inferences about India are not borne out by facts. India may indeed be the exception that proves his theory wrong. On the national scale, except for language, the major diversities in India lack well-defined regional concentrations. Their spatial intermix and overlapping distributions have substantially neutralized their divisive potential. As cleavages in Indian society overlap, conflicting identities find their loyalties divided and their interests interlinked (Dikshit 1975, p. 125). Owing to this, the social cleavage theory of elections, so popular in the West, fails to provide a meaningful understanding of India's electoral politics at the national level.

It is a very different matter in regard to state and local elections, however (Chibber and Petrocik, 1989). Furthermore, the predominantly agrarian economies of the various states and geographically patchy industrial development across the country, have ensured against any formidable interregional conflict on economic issues.. The combined force of these factors has militated against the rise of "ideologically" defined mass parties. This has ensured that political parties aspiring for all-India political stakes remain "aggregative" (*i.e.*, all-inclusive) in nature for a long time. But this does not preclude them from developing into parties of mass mobilization. The examples of the Congress party before the 1969 split, and the present day Bharatiya Janata Party would demonsstrate the validity of this proposition.

Despite the depressed state of its economy, India stands out as a true example of liberal democracy. Reasons for the durability of liberal democracy in India must be looked for in the country's historical antecedents that over the millennia have led to the rise of a common cultural ethos, binding the diverse parts "of the country into an organic entity to override regional differences. India's political nationalism (in contrast to Europe's) was, therefore, an inclusive nationalism (Dikshit, 1981). This fundamental fact of Indian history lies at the root of the all-inclusive nature of the genuinely all-India political parties, none of which can afford to be doctrinaire or divisive.

Unlike many areas of comparable extent in the Old World, India from the beginning developed as a vast cultural realm made up of a number of regional units that had always been conscious of their unity in diversity. The geographical layout of the country had provided favourable circumstances under which

> Hinduism was able to build up throughout India a common culture and a uniform way of life... The foundations thus existed on which, under favourable political conditions, a structure of true unity might be created (Griffiths, 1962, p. 17).

As Spear (1968, p. 110) wrote, the division in the political life of India was not so much a chronic malady (as suggested by many British historians). It was a periodic malaise caused by the nature of Indian polity and the tensions produced by overlapping "races" and "cultures" in a subcontinent which provided few convenient physical compartments for the growth of divisive nationalism. Unity was further reinforced during British colonial rule, when the whole of India had developed as a single economic unit bound together in a common market through an effective communications network. The provinces had no reason d'etre as independent economic organizations, and industrial development had taken place at the most convenient locations disregarding political boundaries.

As such, the liberal democratic federal political system of independent India was not an artificial creation. It represented the emergence of a single nation welded together by social, economic, and political forces reinforcing each other, over thousands of years (Dikshit 1975, pp. 121-122). Herein lies the secret of the durability of liberal democracy in India.

FURTHER COMMENTS ON THE WORLD-SYSTEMS INTERPRETATION OF ELECTIONS

In a recent review of electoral geography, Reynolds (1990) wrote that Taylor's attempt to "integrate electoral geography and political geography" by introducing world-systems concept, "must be judged only as a partial success":

> Part of the reason lies in his implicit treatment of parties as theoretical entities rather than as historical

creations. Political parties are viewed as some sort of complex switching mechanism engaged in what might be called "double duping". On the one hand, parties are the dupes of various fragments of domestic capital—the politics of power—and on the other, they are seen as cynically duping the masses in order to obtain and remain in power—the politics of support.

> Parties may be masters at organizing some issues into politics and others out...but they are unlikely to do so solely because of "structural imperatives"... The more likely reason is to win elections without alienating their traditional supporters. The difficulty with Taylor's "two politics" schema is that it fails to situate political parties in any coherent theoretical and/or historical context....as a theory of parties and elections, it is far too mechanistic and abstract. (Reynolds 1990, p. 26).

Reynolds also criticised Taylor for placing too much emphasis on the role of parties as mobilizers of non-class cleavages and not enough on class conflict in its politics of support. Taylor was also criticized for putting too much emphasis on conflicts internal to the capitalist class in its politics of power. As Reynolds noted, for mass political parties,

> the politics of power...entails the formation of the equivalent of a national coalition consisting of factions and segments of both capital and labour. Although this may occur only in advanced capitalist countries, where class-struggle is sufficiently far advanced for material interests of significant sections of labour to be congruent with those of factions of capital, it is a contigency that is not easily accmmodated within Taylor's original schema.

Another serious deficiency in Taylor's analysis is in that "by retaining the concept of 'politics of support' it continues to bestow ontological privilege to particular 'cleavages' in an essentially national electorate rather than to particular places in a highly regionalized electorate". In other words:

> Taylor's analysis assumes that national parties, in their attempt to win elections, identify issues which appeal to particular segments of the electorate irrespective of

> where such segments may be located. This assumption is a questionable one.... There is an alternative view of electoral politics (and by implication in electoral geography); one in which places provide an essential context in any explanation of social relationships including political behaviour. (Reynolds, 1990, p. 27).

In the context of the nature of contemporary politics under the world-systems perspective on elections, it has been noted that the State is a capitalist State in a dual sense: Firstly, it is part of a larger capitalist system, and second, it pursues policies designed to enhance capital accumulation by those capitalists who control its governmental apparatus. According to Reynolds:

> There are some strong parallels between this perspective on "international" politics and Harvey's recent work on "urban" politics (1985a) and the geopolitics of capitalism (1985b). A basic difference is that in Harvey's theory the ruling coalitions which form to maintain and enhance fixed investments of capital in particular places, are cross-class alliances made up of both factions of capital and segments of labour, whereas in Taylor's world-systems perspective the dominant coalitions controlling States consist of particular factions of "national" capital. (Reynolds, 1990, p. 24).

Reynolds described Harvey's theory as "specific to advanced capitalist States (those in the 'core' of the world- economy)", whereas the world-systems framework is advanced to be applicable to any part of the global world economy.

Once a country has taken to industrialization, class conflict becomes inevitable (even if it is not so pronounced as, say, in Western Europe) as the rise of communist parties ruling over certain parts of India would show. There have been over the last quarter century different combinations and permutations of political forces at the national level wherein the dominant left has been an indirect participant in the central government. This was the case with the Communist Party of India in the 1970s, and with the Communist Party of India (Marxist) during the V.P. Singh regime. Thus, types of cross-class alliances are by no means absent outside the advanced capitalist States. The

Indian National Congress has traditionally been such a coalition, as have also been the other centrist political formations since 1967.

FROM POLITICAL METHODOLOGY TO GEOGRAPHICAL SOCIAL THEORY: THE RENEWED FOCUS ON PLACE-CONTEXTS

As John Agnew (1990, pp. 17-19), noted, one continuity running through much of contemporary electoral geography is the assumption that geography is either epiphenomenal (*i.e.*, the outcome of deeper national, global economic, or political processes) or residual (*i.e.*, it was important at some early stage in historical development but otherwise increasingly unimportant).

> Geographers have made either one or both of these assumptions because the political methodology they have used, based largely around the application of the general linear model to data organized by census categories (regression analysis, factor analysis), has led them to add distance as a variable or to use aggregate census data in a way that denies the pervasive geographical constitution of social processes.
>
> According to Agnew,
>
> The key claim on behalf of a geographical social theory is that national social categories are neither empirically appropriate nor theoretically coherent as causes of voting behaviour. National categories cannot cause individual voting decisions. Causality can only be developed as specific social mechanisms that translate social structure into individual acts and vice versa. From the point of view of this social theory, therefore, electoral geography needs to shift its focus from the psychology of individual voters and their census attributes, to the history of places and parties (Agnew, 1987, p. 18).

"All places have their own history, 'external' links and peculiar traditions. Consequently, places are the social contexts in which political expression is determined" (p. 17).

A second continuity in contemporary electoral geography lies in

> the presumption that empirical findings from application of the dominant methodology somehow "speak for themselves" and make sense because of their origins in that methodology rather than because they make sense theoretically... Something akin to what Primo Levi (1986, 148-9) notes of the practice of industrial chemists afflicts the practice of electoral geographers: with the passing of the years, what had been a crude measuring operation had lost its significance and was transformed into a mysterious and magical practice (Agnew, 1990, pp. 18-19).

Agnew names this continuity as the "faith in the grace of methodology under which "The message determines the medium: the medium delivers the message". Elaborating on the theme of the politics of place, Reynolds (199, pp. 27-31) underlined that throughout the 1960s and 1970s, the standard approach in explaining particular geographies of voting was to relate measures of the social composition in constituencies to some measure of partisan support within them.

> This was the so-called social cleavage model. Departures from the "model" expectations were frequently attributed to unspecified regional "effects" or to unspecified sectional parochialisms. In States like the US where it was well known that the patterning of the relationships varied over large regions, models had to be calibrated on a regional basis. The prevailing view was that such regional efforts, whatever their cause, were relatively minor and soon to be relegated to the "scrap heap of history" as the voting response of the electorate became "nationalized".

As Reynolds put it, electoral geography remains loath to abandon its heavy reliance on a "compositional" approach to the explanation of human behaviour in favour of a more "contextual" approach. In a compositional approach a person's partisan attitudes are viewed as a function of his position within the division of labour. Thus, it is believed that the knowledge about a society's cleavage structure provides sufficient ground for the prediction of voting behaviour.

4

SYSTEMS OF GOVERNMENTAL ORGANIZATION: FOCUS ON FEDERALISM

INTRODUCTION

"The study of federalism belongs to the field of Political Geography"—*N.J.G. Pounds.*

The states of the world are classified on the basis of form of govern meat into unitary states and federal states. In a unitary state there is only one government which exercises its power over all parts of the state. On the other hand, in a federal state, the central (federal) government exercises its power on subjects such as defence, foreign affairs, currency, communication etc. and the governments of component units enjoy the freedom to manage the internal affairs such as education, revenue, justice, law and order etc. In other word, Government comprises a set of legal and political institutions that regulate relationships among the members of a society and outsiders. These institutions have the authority to make decisions for the society on policies affecting the maintenance of law and order, and the achievement of certain societal goals. Modern governments in the democratic system generally derive their legitimacy from national constitutions that provide a legal framework for their rule and specify how power is to be exercised and controlled.

An effective government requires a workable method for distributing authority within the country. This method is contained in the national Constitution, which provides detailed guidelines for the organization of the governmental system. The constitutional instrumentality differs from country to country keeping in view, the great diversity in historical traditions and socio-economic conditions. Broadly speaking, there are two different principles of governmental organization: unitary, and federal.

CONCEPT OF FEDERALISM

It would be worthwhile at the outset to examine the concept of federalism. The term federalism is derived from the word 'federal' which means "of a state in which several states form a unity (union) but remain independent in internal affairs". Federalism is a geopolitical concept and denotes a set of institutions, established for meeting a particular socio-economic and political situation that cannot otherwise be faced in a country of vast and spatially grouped differences.

"The federal principle", writes N.J.G. Pounds (1963), "is a government device whereby outward unity is maintained while allowing a degree of autonomy in some fields of government to the constituent regions of the states". According to Tarlton (1965) federalism is almost universally defined implicitly in terms of component political and legal units sharing equally the federal relationship among themselves and with the general (central) government. Federalism is, thus, stratified system of governance of the state.

It is clear that the essential element of federalism is a division of powers between the federal government and the governments of the component states. For example, in India a clear-cut division of powers between the Union government and State governments has been provided in the Constitution of the country.

The Central governments can legislate on subjects mentioned in the Union List while state governments are empowered to legislate on subjects in the State List. There is, however, a third list, the Concurrent List, on which law can be made by both the governments. The residuary powers lie with the parliament which has the exclusive power to legislate on subjects not included in any of the three Lists.

There is no agreed definition of federal state. Several forms of federal state, therefore, exist in the world today. They range from one tending towards the unitary form (India) to truly federal form (U.S.A.). Federalism can take a variety of forms since all federal states have not originated in a similar manner. Some federal states came into existence because there was no alternative to federal form of government. India can be cited as an example. Some federal states of the world have originated as a result of an agreement, signed by the federating units, which bad the right to leave the federation, if they so desired. ..The foxier Soviet Union serves as an example of sash a federal state.

THE CONCEPT OF SYMMETRY AND ASYMMETRY IN FEDERAL SYSTEM

C.D. Tarlton (1965) considered the relationship of the component states in a federal system with the central (federal) government and with other member states and introduced two concepts—the concept of symmetry and the concept of asymmetry. According to him, "an ideal symmetrical federal system would be one composed of political units, comprised of equal territory and population, similar economic features, climatic conditions, cultural patterns, social groupings and political institutions." Thus, each component state would face similar problems and have similar relationships with the central government. It also implies that tee division of powers between 'the Central and State governments 'would almost be the same in every case. However, since each member state is a separate unit, it would face general problems of its own, the solution of which is the responsibility of the state government. The federal government would mainly solve problems pertaining to whole federal system such as the problems of international relations, defence etc. Tarlton, however, admitted that "no federal arrangement is to bo made up of states, each of which stands in exactly the same relationship to the whole system". In such a federal system complete harmony in centre-state relations may be expected.

An ideal asymmetrical model is the reverse of the symmetrical model. "In the model asymmetrical federal system, each component unit would have about it a unique feature of set of features which would separate in important ways, its interests from those of any other state or system considered as a whole." Such a federal system is composed of unique states, each of which has different problems and competitive interests. As such it is most likely that the federal-state conflict may occur, the resolution of which may sometimes be difficult. The degree of harmony or conflict within a federal system is then a function of the symmetrical or asymmetrical pattern, prevailing within the system. Tarlton opined that "most real federal states, however, could be somewhere between the complete harmony of the symmetrical model and the complete conflict-potential of the asymmetrical model".

FEDERATION, CONFEDERATION AND UNION

The term 'federation' is derived from "federate' meaning to unite together in a league for some common object. It is a group of states, organized on a federal basis. Obviously a federal state and federation are the same. However, there is some difference between federation and confederation, although, the two terms have been used as synonyms in political literature. Switzerland, which is one of the oldest federal states of the world, has been often called confederation. It is obvious that there is little to distinguish between federation and confederation, since each of the two terms signifies a 'contract or agreement among independent states. Pounds (1963'), however, has opined that the confederation is "a term, reserved for federal organizations of a loose order". For him the two terms perhaps mean different entities. He has cited Switzerland as an example of a confederation. The term union has been used in the sense of a federation. 'Union' includes only the states which enjoy the status of being members of the federal system and share a division of power with the union. As such union and federation have the same connotation. The former Soviet Union and Indian Union are the examples.

UNITARY STATES

In countries with relatively homogeneous population having common traditions, language, and sense of national history, the central government ordinarily retains most of the administrative powers. Such a principle of governmental organization is called unitary. In this system the central government is the sole arbiter of policy programmes, and performs all the major governmental functions. Subnational units administer matters within their jurisdiction, but their powers are set and delegated by the national authority. All significant laws are national in scope, and the national government remains the sole taxing authority. It is also the sole custodian of police power—the inherent power of the government to provide for the health, safety, and welfare of the citizens.

Most nations of the world are organized into unitary States, though their institutions and processes present a great variety. In Great Britain, for example, a certain degree of regional authority exists in Northern Ireland, Scotland, and Wales, and the local county governments perform certain fairly autonomous functions. This is sharply contrasted by the French example, where strict control over the administrative territorial divisions is exercised by the national government, so that there is little by way of regional government.

In countries with heterogeneous populations, especially where the component ethnic identities are regionally concentrated, or where there is a sharp regional division in economic interests, if the component regions possess conflicting historical traditions, a unitary system of government shall be difficult to operate.

In such cases the preferred form of government is some kind of federation. Under the federal system, governmental authority is "layered" among different levels. The rest of this chapter fucuses on the nature, origin, and stablity of federalism as a governmental system.

THE NATURE OF FEDERALISM

Although Federalism is so much in currency and the federal form of government by no means uncommon, the term "federalism" is not easy to define. "Devised as a form of constitutional government to express imperfect unity and multinationalism, federalism is a particularly complicated form of western democracy" (Camel 1961, p. 16). The term "federal government" is often used very loosely in political discussions and is seldom given a meaning that is at once clear and distinct.

The main features of a federal system are, however, well-known. *A federation is born when a number of usually separate or autonomous political units (or units with some pretensions to autonomy) mutually agree to merge together to create a State with a single sovereign central government, while retaining for themselves some degree of guaranteed regional autonomy*. The merger of the units in a federation is not absolute but partial, and its degree may vary depending upon the circumstances of the particular groups of the political communities involved. The legislative and executive powers in a federation are divided in a coordinated fashion between the federal (*i.e.*, central) and the unit governments, each of which acts directly on the people—the central government having jurisdiction over all matters, that have bearing on the development and security of the nation as a whole, and the unit governments with their right to regulate the matters of local and more immediate importance to their respective peoples.

Federalism is, therefore, essentially a compact. Like other compacts, it has a written constitution that cannot be unilaterally altered. The terms of the compact, and the division of powers or functions therein, are made by the federating units as coordinate constitutional bodies, and not by a dictatorial third party or an overbearing unit within the group. To ensure that no undue and unauthorized inroads are made by one level of government into the sphere of the other, there is usually a judicial review by a supreme court acting as the ark of the federal covenant. Moreover, as "money is....the vital principle of the body politic—which sustains its life and motion and enables it to perform its most essential functions" (Hamilton, Madison and Jay, 1961 edn. p. 188), in order that the federal government and the units are truly coordinate in authority, it is necessary that each has a good measure of control over its finances, and, usually, an access to taxing powers.

It should, however, be emphasized that federalism is a set of institutions erected to serve a particular set of social, political, and economic situation. The phenomenon that is so created is not static but dynamic. It goes through a process of evolution and change, because a complex of psychological, social, political, and economic factors which necessitate federalism may require one type of instrumentality at one time and quite a different one at some later date. In fact,

> as the nature of the society changes, demands for new instrumentalities are created and these demands are met by changing or abolishing old instrumentalities and establishing new ones in their place. (Livingston, 1952)

Federation, Confederation and Unitary States

Federalism differs from certain other forms of government. The two words "federation" and "confederation" have often been used as synonyms even by serious students of constitutions in the past. *The Federalist* itself did not distinguish between the two terms; and even Dicey used the two words rather loosely when he wrote: "The Physical contiguity... of countries which are to form a *confederated* State is certainly a favourable...condition for the success of federal government" (Dicey, 1939 edn., p. 603). And etymologically, there is little to distinguish between federal and confederal, for each of the two terms implies a covenant, compact, or treaty among independent States. The oldest meaning of the expression federal government appears in its use to refer to loose linking together by a treaty of sovereign States for specific military or economic purposes. Examples of this type are found as far back in history as the confederacies of ancient Greece.

Modern scholarship has, however, insisted on drawing a clear-cut distinction between the two terms, despite the fact that such federal States as Canada and Switzerland describe themselves as "confederation (s)". As Wheare wrote, a confederation is now described as "that form of association between States in which the general government is dependent upon the regional governments". It is represented by countries whose constitutions "embody the principle of subordination by the general government to regional governments". (Wheare 1963, p. 32) Thus, a federation differs from a con- federation in the sense that in the latter the central government is subordinate to the unit governments, in the sense that it functions at the mercy of the regional governments. But in a federation neither level of government is at the mercy of the other.

In a confederation there is no direct contact between the peoples of the constituent units and the central authority. The central authority in a confederation is compelled to reach the people only through the respective regional governments, which may or may not allow this contact. In a federation, by contrast, there is direct relationship between the central government and the people, who not only have a share in the task of constituting it but also submit to its rules (in the spheres of its competence) without the interposition of the regional governments as intermediaries (Hughes, 1963, p. 13).

In a confederation, the member States retain their sovereignties and, therefore, the central authority cannot compel its decisions on any of the constituents. The central authority in a confederation can act only when all the constituents are unanimous. As The Federalist recorded: "In our case, the concurrence of thirteen sovereign wills is requisite, under the Confederation, to the complete execution of every important measure that proceeds from the union" (The Federalist, essay no. 15). But in a federation there is no division of sovereignty, the constituent units are only autonomous in certain specified spheres. Once a federation is created, the States have to abide fey the decisions of the properly constituted central government in matters where the constitutional compact has empowered it to act. Thus, as MacMahon wrote:

> The logical difficulty of divided sovereignty can be avoided... by regarding a confederation as a merely, comprehensive and cohesive form of international administrative union, whereas a federal system is regarded as a multiple government in a single State. (MacMahon, 1931, p. 173)

A federation differs from a unitary government in the sense that in a unitary polity "states", if any, exist at the mercy of the central government. In a federation on the other hand each level of government is, in theory, autonomous within its allocated sphere of competence, and is free from any non-agreed intervention from the other, except in an emergency, if the constitution so provides. Thus what distinguishes federalism from a unitary government is guaranted constitutional autonomy, not the formal division of powers.

Federalism as a Constitutional Compromise or Bargain

As Dicey pointed out, federalism rests on the peculiar psychology of the peoples of the different federating units, of desiring union without desiring complete unity (Dicey, 1939, p. 602). A federation is born only when the political units in a region possesses some very strong factors of individual identities which create in them a genuine desire to maintain separate existence, while at the same time, there are certain factors of vital import that all these units share in common and, for which matter, they desire a strongly coordinated and united existence. Faced with such a dilemma, when they can neither set up separate houses of their own, without losing the far greater advantages from union, nor amalgamate into a complete union without forgoing their individual identities which they so greatly value, the political units enter into a compromise arrangement to create a halfway house between complete unity and complete separation, and a federation results.

Federalism is, thus, essentially a compromise between centripetal and centrifugal forces that are operative at the same time. It is born only when a characteristic balance between these forces is reached. It is clear that because federalism is essentially a bargain, the units would merge into a federation only when the centripetal forces, in some ways, overwhelm the separatist ones, and the units in general see greater advantages in union than in separation (Riker, 1964). Still, the basic problem of a federalism has traditionally been for the federating units "to keep the centrifugal and centripetal forces in equilibrium so that neither the planet States shall fly off into space nor the sun of the central government draw them into its consuming fire" (Bryce, 1988). Erected essentially as a halfway house between complete unity and complete separation, federalism clearly has a wide spectrum (Livingston, 1956, p. 4).

Stages of Federalism

During its pre-twentieth century phase, federalism was more or less a dualistic polity, "in which the federal and the State governments pursued virtually independent courses of action during a period when government activity was in any case minimal" (Elazar, 1964). Federalism then consisted of "two separate federal and State streams flowing in distinct but closely parallel channels" (Clark, 1938). Describing the American federalism in 1858, Chief Justice Taney observed that:

> The powers of the general government and the State, although both exist and are exercised within the same territorial limits, are yet separate and distinct sovereignties, acting separately and independently of each other, within their respective spheres (Cited in Elazar, 1964).

This traditional dualistic approach to federalism was expounded by scholars like Freeman (1893), Dicey (1885), and Garran (1901), and has in more recent years been refined and justified by Wheare. This legal theory of divided sovereignty and the two distinct and separate spheres fitted the facts of the time well enough though not perfectly, for till long after the Civil War in the United States the few activities of the national government could go alongside the limited state activities, without either impinging seriously on the other. It was almost, if not quite, a "functionless federalism" when compared with the present-day conditions (Anderson 1946, p. 13). But the economic philosophy on which this dualistic federalism was based, has become quite outmoded in the present context. No modern State, irrespective of its economic and political philosophy, can now avoid an extensive degree of State intervention.

This is an era of the active, public-service State, not that of the negative Saissez-faire State. The performance of functions and services is the keynote of modern government, and in that performance cooperation, interdependence, interpenetration of national and state agencies are inevitable (Anderson, 1946, p. 14). For this reason, older constitutions have been adapted to fit the needs of the present time by the development of extra-constitutional devices such as administrative cooperation between governments, coordination of state policies by conditional grants from the federal government, and the purchase of federal monopoly or near-monopoly of taxation of incomes and profits. Many scholars, therefore, think that federalism has become obsolete in the twentieth century for, as Karl Loewenstein (1951, p. 212) believed, economic planning is the DDT of federalism. But federalism has only entered a new phase. In fact, whereas the guiding principle of the eighteenth and nineteenth century federalism was the independence of the state and federal authorities, the guiding principle of the mid-twentieth century federalism is the need for cooperation between them (Birch, 1955, p. 305). It is only the pest of dualism that economic planning (the DDT) has killed. It is the dualistic phase of federalism that has now become a relic of the horse and buggy days.

This new phase has been called cooperative federalism which, in essence, is a system by which state and national governments supplement each other and jointly perform a variety of functions. The national government, in the new phase, with its greatly enlarged powers and functions has only "supplemented rather than supplanted the performance of functions by the States" (Livingston 1968, p. 132). The philosophy of the earlier federalism was, as Grover Cleveland (President of the United States, 1884-1888, and 1892-1896) is reported to have said, that, it is the duty of the people to support the government and not that of the government to support the people. Now the philosophy has greatly changed. "There is no longer any question about the national government's power to act, but only about the appropriate means and amounts and the proper timing of the actions to be taken" (Anderson, 1960, p. 12).

As some recent studies show, the traditional picture of the nineteenth century American federalism is unreal, and federalism in the United States, in practice, if not in theory, has traditionally been cooperative (Vile, 1961, and Elazar 1962). The theory of dual federalism was not viable when applied to concrete problems in specific situations even in the early days of the American Republic, wrote Elazar, and added that federalism when interpreted to mean demarcation of responsibilities and functions has never worked in practice. While the amount of governmental activity on all planes in relation to total activity of the American society has increased, the governmental activity that existed in the nineteenth century was shared in much the same manner as governmental activity in the twentieth century. Indeed, the roots of cooperative federalism are entwined with the roots of federalism itself.

In view of the changed emphasis on State and federal government co-operation, Wheare's insistence that each of the two levels of government should be limited to its own sphere and within that sphere should be independent of the other, appears excessive. Birch, therefore, proposed to delete these clauses in Wheare's definition of federalism. As Birch put it,

> A federal system of government is one in which there is a division of powers between one general and several regional authorities, each of which, in its own sphere, is coordinate with the others, and each of which acts directly on the people through its own administrative agencies (Birch, 1955, p. 306).

This avoids confusion regarding federalism and quasi-federal ism. Approached thus, federalism will not appear "obsolete" (Laski, 1939, p. 367). Wherever the problem of securing political unity in the face of regional diversity is to be reconciled in future, federalism would prove adaptable enough to continue to serve a fruitful purpose.

Wheare found the above modification of his definition as one of the most constructive and fruitful, and most students of modern federalism would agree that on the whole, this modified definition provides one of "the most serviceable definition(s) of modern federalism" (Wheare 1963, p. 14). In fact, "Under the heat and pressure generated by social and economic change in the twentieth-century, the distinct strata of the older federalism have begun to melt and flow into one another" (Corry, 1958, p. 122).

The following excerpts from some recent studies would further clarify the concept of modern federalism. Morton Grodzins stated:

> The American form of government is often, hut erroneously, symbolized by a three-layer cake. A far more correct image is the rainbow or marble cake, characterized by an inseparable mingling of differently coloured ingredients, the colours appearing in vertical and diagonal strands and unexpected whirls. As colors

> are mixed in a marble cake, so functions are mixed in the American federal system (Grodzins, 1960, p. 265).

Thus, the nineteenth-century brand of primarily legalistic and dualistic phase of federalism is over. Federalism is no longer "like a great factory wherein two sets of machinery are at work, their evolving wheels apparently intermixed, their bands crossing one another, yet each doing its own work without touching or hampering the other" (Bryce, 1988, p. 432).

As William Anderson (1946, p. 3) wrote:

> the entire network or structure of American [so also other federal] governmental units—national. State, and local—has become so close-meshed in recent decades, that a strain or change at any point has repercussions in other parts of the fabric. One part cannot be understood if separated from others.

Regional and central governments in a federation, at the present, should not only be coordinate but they should also be cooperative. Modern federalism is, therefore, basically a federalism of function rather than of powers—a federalism more of politics than of laws. It is important to remember, however, that the difference between the two versions of the federal concept is chiefly one of emphasis. Whereas "dual federalism" viewed the two sets of government primarily as equal rivals, "cooperative federalism" views them as equal partners. (Even though the central- government tends to become more equal than the state governments). What lies at the root of both, the views is the premise that in a federation (in principle) neither level of government is subordinate to the other (Watts, 1966, p. 13).

Although the States-rights sentiment continues to exist in some form or the other in almost every federal State, the nature and degree of rivalry between the constituent states and the central government that existed in the dualistic phase of federalism has undergone drastic change. Through a long process of adjustment, the main areas of state and central functions are largely agreed upon (though disputes by no means are uncommon). The main rivalry now is between the states themselves, wherein the centralized regulator, *i.e.*, the federal government, plays the fundamental coordinating role. This is so because "different regions tend to regard themselves as rivals just as much as small nations are within a common market" (Boudeville, 1966, p. 57). The States also compete one another for grants from the central government.

The increasing role of the federal government as the centralized regulator brings us to a new phase in the development of federalism. This new phase has been termed as organic federalism, but in view of the discredit associated the word "organic" in political geography, it may be termed as integrated federalism. But as Sawer wrote, this is, still a recent and ill-defined concept. Organic federalism is a category of federalism in which the Centre wields such extensive powers and gives such a strong lead to the state governments in some of the most important areas of their individual as well as their co-operative activities, that the political taxonomist might hesitate to call it federal at all. Sawer (1969, p. 125) regarded Austria as the most obvious candidate for the organic category, because here although the "Centre dominates every aspect of policy", "Region autonomy within the limits of the Region competence is no sham, and the values inherent in such autonomy are protected both by the constitutional structure and by the pattern of politics". Sawer thought that among the older federations, the United States is, the only one where a surge towards organic federalism could take place in the near future, somewhat similar to the surge towards cooperative federalism in the late 1930s.

The question that arises now is, What is the line of demarcation between organic federalism and the organic decentralized unitary State? The answer is not very difficult. The essence of federalism is that each level of government should have guaranteed autonomy. Thus,

> so long as the amending procedure, the operation of the judicial review and the pattern of politics or a combination of any two of them restrict the ability of the Centre to abolish a Region structure...the position of a Region is sufficiently secured, and the polity in question should be called federal. (Sawer 1969, pp. 127-128).

THE GEOGRAPHY OF FEDERALISM

Although many would now regard federalism as the most geographically expressive of all political systems" (Robinson, 1960), so far, scant attention has been paid by geographers (or any one else) to explain the geographical basis of the federal polity. Livingston (1952) who revived the concept of federalism as a polity based essentially on regionally grouped diversities—and to whose writings much of the recent interest in federalism by geographers is due—himself described the concept as sociological rather than geographical.' The result is that students of federalism have often wondered whether Livingston had, indeed, provided us with "a useful tool of analysis" (Birch, 1966, p. 15-33).

There are two reasons why federalism is considered to be the most geographically expressive of all forms of government. First, it is based on the existence of regional differences or a sense of locality, "the belief that the area in which one lives is different from other areas, even though contiguity with them may provide many interests in common" (Miller, 1964, p. 138). Federalism has been described as "the process by which a widening sense of social [and political] solidarity is reconciled with the attachment for local identity, through the provision of dual political organization" (Karve 1932, p. 8). Secondly, because of a sort of dual political organization and the grant of substantial regional autonomy, the constituent regions in a federal State remain highly articulate. Owing to this a spatial interactions in a federal system of government, unlike in a unitary one, are most clearly recognized. Because federalism starts with a tacit recognition of the immutability of regional personalities, and because inter-regional interaction in the political and economic life of the federal States are most clearly recognized, federalism becomes a most apposite subject for geographical inquiry.

Since the basic geographical premise of federalism is the existence of regionally grouped diversities, "no government has ever been called federal that has been organized on any other but a regional basis....federalism becomes nothing if it is held to embrace diversities that are not territorially grouped" (Livingston, 1956, p. 3). It is true that regional differences, or a strong sense of locality may sometimes exist even in States that are not organized on a federal basis. But the fact to be remembered here is that only when the sociological unit of the region is powerful enough to demand and receive social accounts does federalism become inevitable. The regional differences that, for example, exist in France have so far not been strong enough in this respect. France, therefore, remains organized on a unitary basis. Similarly, regions that were once very diverse and possessed strong regional identities may currently be found to be joined together under a single unitary State where, even though regional identities continue to be present, through a long process of adjustment the cultural-sociological unit of the region has ceased to demand account. This often happens when the regions concerned were brought together by imperial conquest, so that the region or the regions in question were not in a position to assert the recognition of their special position.

Sometimes, even clearly identifiable regions in a unitary State are not able to demand social account in the manner that regions in a federal country are able to. This is because, in a federal State, regional identities are respected and protected by the Constitution which guarantees the claim of each component unit of the State to perpetuate its identity. In a unitary system, to the contrary, these differences and diversities are largely suppressed or ignored. It needs clarification, however, that federalism does not encourage the perpetuation of sovereignties if there be any. The fundamental basis of a federal union is a clear recognition of the limitations of the constituent units as self-sufficient and completely functional entities.

Federalism, unlike a unitary system, does not force unity out of diversity. Through the system of guaranteed autonomy to the component Regions, it allows the two to coexist. But in the process of its progress towards maturity, it does create unity through the progressively enlarged functions of the central government (national planning and the like) and also through the falling down of the once rigid barriers (physical, psychological, and economic) between the component units of the federation. Confusion of this nature will be avoided if we remember that, like all things human, federalism is also constantly in a state of flux. It is not a static phenomenon but a dynamic process. Federalism is, in some senses, truly a halfway house to unity and integration, though not to a unitary form of State. A federation once established, tends to rigidify existing regional identities by giving them continued opportunities for articulation. Even though the central and the regional governments become largely cooperative, the regional governments continue to remain as rivals. Thus, "the antithesis of federalism is not unitarism, but particularism and separatism" of an extreme kind (Boehm, 1931, p. 170).

The preceding discussion showed that regionally grouped diversities are the fundamental fact of federalism. It needs clarification, however, that the geographical distribution of diversities within the State need not strictly follow the boundary lines of the constituent units of the federation. Since federalism often embraces a series of diversities on a number of issues, it can not be expected that the inter-regional boundary lines will mark off areas which coincide with all the different interests and opinions which may be held on all possible questions. The essential fact to remember is that the units in question should possess a total complex of diversities that are strong enough to distinguish them from their fellow members in the group and, thereby, make them demand and receive recognition of their individual identities. Regionalism of this kind, in which diversities spill over state boundaries, is considered a valid manifestation of the federal principle. My study (Dikshit 1975) of the Swiss and other federations shows that it is also a beneficial manifestation.

Diversities within a federal society may turn on all sorts of questions—economic, religious, historical, linguistic and cultural. Any of these, or any combination of these, may produce a group demand for self-expression. The major diversities exerting major influences on national politics may have two contrasting patterns of geographical distribution. They can either be territorially or regionally grouped, or they may be mixed together and scattered like the strands of different colours in a marble cake. If the major, identifying diversities within a State are arranged territorially, then the society or the political community is potentially federal. But if the distribution of diversities follow a marble cake pattern, the society is plural and non-federal. Only when a society contains territorially concentrated groups so markedly different from one another that they must insist upon some instrumentality to protect and articulate their special identities, does the need for federalism genuinely arise. Doubtless, "one such circumstance...does not make the society or constitution federal. But two or six or twenty may produce a result that may properly be so called" (Livingston, 1956, pp. 2-3).

A word of explanation is necessary lest the above statement regarding the geographical pattern of diversities creates confusion. Here we are concerned with the distribution of diversities and not with functions (that Morton Grodzin's statement in the first part of this chapter referred to). The distribution of functions in modern federations may often resemble the marble cake pattern, the major diversities on which a federation is based, must essentially be territorially arranged, otherwise the society cannot be called federal. This does not contradict the pattern of diversities exemplified by Switzerland, where each of the two major diversities in the nation's life is territorially arranged, though one set cuts across the other and thereby has contributed to create a unifying factor in the functioning of the federation (Fig. 4.1).

For long, federalism remained a subject matter for purely legal discussion. But legal answers are of value only in solutions to the purely legal problems. Federalism is concerned with many problems other than the legal ones. Hence, a purely legal approach to federalism has not sufficed. As Livingston (1956, p. 1) wrote:

> The essential nature of federalism is to be sought for, not in the shadings of legal or constitutional terminology, but in forces—economic, social, political and cultural—that have made the outward form of federalism necessary.

Like most other institutions of man, federalism is an attempt to resolve a certain set of problems of human organization. The particular task of federalism, is to find solutions to governmental questions involving complex interaction of inter-regional differences and similarities. Federalism is, therefore, essentially a product of the geography of the constituent regions. In this sense, its essence lies not so much in the constitutional structure as in the geography of the society itself.

This geographical view of the nature of federalism, should do away with the confusion that over-emphasis on the "sociological" view of federalism frequently created. Sawer (1967, p. 136) objected to Livingston's statement that federalism is "a function not of constitutions but of societies". Such statements, Sawer wrote,

> can be misleading, because they suggest that there is a sort of general social attitude, or type of social structure, which corresponds uniquely with the constitutional form known as federalism....I do not believe that these attitudes or structures are specific to federalism—the favourable social attitude is an attitude towards government, administration and law in general, not towards federalism as such. (Sawer 1969, p. 136).

Unlike the social structures and attitudes, the geography of the society (as underlined above) is to a very large extent, unique and specific to federalism. But the geography of the society, in no absolute sense, determines the federal or non-federal form of government that may ultimately evolve. Sometimes federations created under or enforced over largely nonfederal situations may survive as in the case of Germany after World War II. But such federations survive largely because during the long period that the federation remains enforced, the national population has got adjusted to the changed political situation that it could not undo. The functioning of the Regional governments as autonomous entities creates a new set of vested interests around them, so that the effective political geography of the region gets transformed, making it more suitable to federalism. But federalism created under non-federal situations, unless enforced by external forces, can hardly develop into anything but the federalized unitary State of the type that Austria represents which, as Sawer said, the political taxonomist may hesitate to call federal at all. On the contrary, if a unitary State structure is imposed over a region that is essentially federal in its politico-geographic structure, the government can be run only by military dictatorship as in the case of Burma or Pakistan, or by one-party rule (*i.e.* group dictatorship) as was the case in the erstwhile Soviet Union or Yugoslavia.

As a particularly dynamic and complex phenomenon, federalism is, in the words of Watts (1966, p. 16), of interest not only to the constitutional lawyer concerned with the nature of legal frameworks, and to students of political institutions, occupied with the operation of particular types of political institutions; but also to the sociologist, the economist, the geographer, the historian and the poetical theorist. Although of special relevance to politico-geographic study, federalism may be found to be of considerable interest to the social geographer interested in the phenomena of social integration and diversity, the economic geographer engaged in the study of the role of political institutions in fostering or hindering economic growth, and the historical geographer interested in the study of the role of spatial interactions in the genesis and evolution of some of the more important "new" nations of today.

Although Livingston drew attention to a fundamental feature of federalism, without a proper appreciation of which the nature and dynamics of the federal polity cannot be fully understood, the issue remained confused and the utility of the concept greatly reduced, because Livingston had described federalism as a "sociological phenomenon". Livingston wrote that "federation is not an absolute but a relative term" and added that the "federal polity is a function not of constitutions, but of societies". Such statements, and the impression that Livingston was proposing a "sociological approach" to federalism, contributed to confuse many a scholar like Birch, who wrote that though it is "true that federal institutions are frequently...a reflection of social diversity, [but] virtually all writers on the subject have said as much. The point is not whether it is true but whether it is useful as a tool of analysis" (Birch, 1966, p. 17).

It is, therefore, important to examine whether or not the concept of regionally grouped diversities as a fundamental premise of federalism is useful as a tool of analysis. Though Livingston's "Note" is said to have given rise to a "sociological approach to federalism", no such approach has ever been defined, nor the concept ever used as a tool of analysis in any study of significance. Whenever the phrase "sociological approach" had been mentioned, it was generally meant to imply that certain sociological factors (*i.e.* social values and structures) are all important in the dynamics of federalism.

So much so that, in his article, Birch did not even once mention the point about federalism as a system based on regionally grouped diversities, while discussing and dismissing the so-called sociological approach.

Birch maintained that there are at least three reasons for doubting the value of the concept put forth by Livingston:

> First, it is clearly very difficult to make firm generalizations about the members of a category of which the membership is undefined...Second, we note that Livingston is forced to include a wide variety of phenomena under his heading "federal instrumentalities", including laws, regulation, theories, concepts, attitudes. Propositions based upon this use of language are bound to be somewhat cloudy: they may not be invalid but they are unlikely to be specific enough to be helpful...Third..., in the main body of his book Livingston is unable to use the conceptual framework which is outlined in the first chapter.

Thus, Birch agreed that the propositions on the nature of federalism that Livingston put forward are not invalid. His main objection is that the propositions were not helpful in the analysis of federalism as a system.

The root of the confusion seems to lie in the failure to appreciate the fact that the essence of Livingston's analysis does not lie in the proposition that federalism is based on sociological diversities per se. The essential feature of the concept is that federalism is distinguished from other forms of government, which may also incorporate a great variety of social, cultural, economic and other diversities by the fact that while in a truly federal society the major diversities are regionally grouped, in nonfederal societies most of the more important diversities permeate the society as a whole without any marked regional concentration. The essential nature of federalism, therefore, does not lie in the societies or the social diversities as such. The essence lies in the spatial pattern of the total complex of social, political, economic and other relevant diversities which have contributed to impart to the regional political communities a definite sense of regional identity.

In the study of federalism as a form of governmental organization, our primary concern is not with the societies or social groups as such, but with the larger political regions (the States) which may themselves be divided into regionally based conflicting social groups. Hence, in order that this concept can prove useful as a research tool, the methodology called for is not sociological, but spatial or geographical. The focus in the study of federalism is not on the component societies viewed as social organizations, but on the states or regions that constitute the federation as a political system. Thus, in order that the concept of federalism as a polity based on regionally grouped diversities could prove a useful tool for research, it is necessary to adopt a spatial-interactional (*i.e.*, geographical) rather than a sociological perspective. As Ullman wrote:

> the main contribution of the geographer is his concern with space and spatial interrelation....Sociology has often been defined by some as the study of social interaction; by the same token geography might be defined as the study of spatial interaction (Ullman 1953, p. 56).

And as Ullman clarifed, by "spatial interaction" is meant "actual, meaningful human relations between areas of the earth surface, such as the reciprocal relations and flows of all kinds...not static location as indicated by latitude, longitude, type of climate etcetra...."

It must be emphasized that a spatial interactional analysis or a geographical approach to federalism (or to any other problem) is not anti-sociological, anti-economic or anti-political. The spatial interactional perspective is primarily a matter of emphasis and mental structuring. Most social sciences deal with the problem of "human organization". But as an extremely complex concept, "human organization" has at least four dimensions which are associated with the evolution of the various social science disciplines. The economic dimension is largely the preserve of the economist; the social dimension is the particular concern of the sociologist and the allied behavioural sciences and the political dimension is largely the province of the political scientist. The fourth dimension is the spatial dimension, which is the particular concern of the geographer. However, the spatial dimension is not the geographers' exclusive domain. "But what we often call 'spatial organization' lies very much at the heart of most geographical investigation, whereas it is [only] peripheral to [the main thrusts of the other] fields.... Whereas the other dimensions usually run parallel to each other in conceptual terms despite various overlaps and conflicts of interest among them, the spatial dimension cuts dead across all others" (Ginsburg, 1969). Indeed, many of the more challenging questions of spatial organization contain elements of all three of the other dimensions. Thus, a geographical approach to federalism is, by nature, complementary rather than contradictory to a sociological, political or economic analysis. As Ginsburg wrote,

> concern with regions of whatever sort usually means a focus on an areal aggregate, rather than upon certain elements of that aggregate. It is here that geographic methodology... [by which word is meant] method in the broadest sense—not merely technique—has a genuine contribution to make to all the social sciences.
>
> As Hartshorne Pointed out
>
> Many fields in the natural and social sciences study a particular category of phenomena, not excluding its distribution and variations over the earth. What geography, and geography alone, studies is the areal character [or spatial organization] of the earth in which man lives—the form, the content, and the function of each areal part, region, or place and the pattern of interactions between areal parts... (Hartshome, 1968, pp. 115-116).

Among the social sciences geography, like history, overlaps the fields which study a particular category—economic, political, or sociological. In geography, as in history, it is the integrated combination of diverse elements in their complex interrelationships, that form the direct subject of study.

Since federalism is essentially a polity based on interactions and interrelations of a complex set of regionally grouped diversities (represented by and articulated through the constituent states of the nation), a geographical approach to its study should prove highly rewarding. Since the chief federal diversities are regionally organized, and since they are neither purely sociological, economic, or political in nature, but an amalgamation of all of them put together, a geographical approach, with its emphasis on areal integrations and regional interactions, should prove helpful in the study of the dynamics of federal political partnerships. The regionally grouped character of the complex of diversities on which federal systems are based, is the most critical factor that distinguish federalism from other governmental systems.

As noted earlier, a fundamental objection that Birch raised against the approach to federalism formulated by Livingston was that Livingston had included such a wide variety of phenomena under the heading of "federal instrumentalities" that "they are unlikely to be specific enough to be helpful". It must be noted here that in geographical analysis the range and variety of phenomena is not significant. What is significant is that regional composition of the complex of diversities should impart the constituent regional units some special regional colour or identity. So long as these diversities are significant in evolving certain specific patterns of inter-regional (*i.e.*, inter-State) relationships, they remain of considerable value in unraveling the dynamics of federalism.

Since the rest of this chapter is devoted to theoretical discussion on the origins and stability of federalism as a political system, a brief statement on how this problem may be analyzed, would appear to be necessary.

It is relevant to recall here Dicey's dictum that federalism results from a peculiar psychology on the part of the federating communities of desiring union without desiring complete unity. Thus, regional loyalties side by side a strong sentiment for national solidarity is the essence of a genuinely federal situation. Therefore, in order to arrive at meaningful hypotheses, we must take a comprehensive view of each federation at the time of its origin, since "any specific arrangement of earth-space becomes fully intelligible only when it is linked with preceding combinations of nature and culture in the same area"

(Whittlesey, 1945). It is imperative, therefore, to examine the historical evolution of society and culture in each of the regional units involved in the federation and reconstruct their economic, political and social geography in the period immediately preceding their union. In this way one can isolate the factors and relationships that were, in each case responsible for creating that peculiar federal psychology of desiring union without desiring complete unity. This should also help in identifying factors that finally overwhelmed the divisive sentiments pulling them apart, and helped finally to bring the diverse units together into a single functioning structure. In reviewing the historical development of each individual federation, our purpose has been, first, to delineate the politico-geographical factors that were responsible for the rise of strong regional identities and the consequent desire for separate existence. Secondly, to isolate the factors which, in the end, overwhelmed their sentiments for separatism, and compelled the constituent units to unite into a centralized federation. From the two sets of factors the researcher can proceed to infer the nature of inter-regional relationships that led to the sentiment for federalism in each case. He-may therefrom arrive at broad based hypotheses on the origins of federalism. With a view to analyzing federal stability, the student should keep in perspective, the working and dynamics of each individual federation since its inauguration, and try to identify therefrom the patterns of socio-political and spatial interactions leading to the success or otherwise of the partnership in each case. The sum total of these conclusions on the rise and survival of federations, past and present, should yield a set of fruitful hypotheses on the origins and stability of federalism as a political system.

A full scale analysis of all federations, past and present, on the lines suggested above (Dikshit

TOWARD A THEORY OF THE ORIGINS OF FEDERALISM

Factors for Overall Union

Enumerated below are some of the more important factors for unity and separation that appear to have played a significant role in some case or other of the federations so far experimented with, even though it is clear that none of these factors, by itself, is either sufficient or necessary for the rise of federalism. Taking the factors for unity first, some of the following factors (in various combinations) have played vital roles in pulling the diverse units together into a union.

A.1. *A sense of common military insecurity and the consequent need for common defence.* This may often be combined with the desire for independence from a common imperial power, and the realization that only through union is this independence possible. Both these conditions had been found together in most cases, *e.g.*, the United States, Australia, India, Switzerland (in its earlier phases), and Malaysia. But only the latter condition was present in post-1945 Austria and West Germany—where the sentiment for union far outweighed the sentiment for regional separation. In the West Indies, neither of these conditions was present, though earlier it was the second factor that had attracted the islands towards a federal union. Both these factors have been treated together, firstly, because federalism (and for that matter any type of union) is essentially a response to the need for common effort. Secondly, when both these factors are present together the sense of unity and common purpose is pronouncedly stronger. A point to be noted is that the second factor, in the absence of the first, may become only a temporary and largely impotent factor for the rise of a durable union. The West Indies and the Central African examples would prove this point.

A.2. *A community of "race", language, and culture (including religion).* This is a very important factor in bringing about that sense of belonging together giving rise to the feeling of community and common destiny in relation to outsiders. In fact, Mill (1947, p. 361) considered this factor so important, as to underline that

> free institutions are next to impossible in a country made up of different nationalities. Among a people without fellow feeling, especially if they read and speak different languages, a united public opinion necessary to the working of representative government, cannot exist.

Examples of multi-ethnic federations such as Switzerland, Canada, or India would show, however, that these factors are neither necessary nor sufficient conditions for the rise of federalism. The fact remains, however, that if the sum total other factors favours the rise of federalism, then community of language and culture shall most likely lead to a' more smoothly-running federal partnership.

A.3. *Interdependent economies and expectations of greater economic benefits from union.* This factor is found to have been present in most examples of federalism leading to stable partnership. Similarly, in most cases where the federal ties have been disrupted or have been threatened with collapse, this factor had invariably been absent, at least in the perception of the dissenting party. It was clearly absent in the West Indies in so far as the largest partner—Jamaica—was concerned, and, the lack of economic incentives was one of the prime reasons why that federation ultimately failed. This factor was also the main cause of Biafran disenchantment in the Nigerian federation, and was the main reason behind the East Bengali disaffection in the Pakistani union.

A.4. *Prior history of the communities as parts of a single overall political-administrative set-up or common experience as parts of the same political system.* This has Been true of every federation so far experimented with, and it may well be a necessary condition for its rise, though whether it is necessarily so or not, is not vital to our discussion.

A.5. *Easy interregional transport and communication or a close geographical neighbourhood.* "Physical contiguity...of countries which are to form a confederated (sic) State is certainly a favourable, and possibly, a necessary condition for the success of federal government," wrote Dicey (1885, p. 603). Though facts prove that this factor is by no means an essential factor in the origin of federalism, but the examples of pre-1971 Pakistan, and the West Indies would show that the absence of this factor is often a source of weakness. The point is partly proved also by the Western Australian example in the Australian federation.

A.6. *The political communities aspiring for federation should possess an overall compatibility in their social and political institutions.* Incompatibility in these matters would often prove fatal, as was proved by the question of slavery in the United States.

Factors Giving Rise to the Desire for Regional Autonomy

B.1. Perhaps the most important factor for the sentiment for regional autonomy is the *existence of regionally-grouped* diversities in respect of

(a) language,

(b) religion, and

(c) 'race',

which may generally be identifiable with the individual political units involved in the federation.

B.2. *Existence of units with independent or competitive economies.* The existence of this factor reduces the scope for economic expectations from the union, and leads to indifference among the units with regard to union. The ill-fated federations of Pakistan and the West Indies present salient examples of the two different types.

B. 3. *Prior existence of the political units as*

(a) independent entities or as

(b) regionally autonomous provinces of a broader political system. It is, however, important that this autonomy should be based either on

(i) a popular (even if restricted) democracy or on

(ii) popularly recognized hereditary kingship, etc., which in the course of its long existence had led to a strong sense of regional identity.

Examples of the first type were presented by the United States, Canada, Australia, and Switzerland, while Malaysia offered a weak example of the second type.

B.4. *Absence of common external threat to the survival of the individual units.* This becomes a critical factor in cases where the aspiring units possess strong (and sometimes exclusive) regional identities born out of their peculiar ethnic, economic or physical circumstances. The West Indies provide a most salient example.

B.5. *Presence of conflicting threats to the individual units.* As Mill (1947, pp. 366-367) wrote, a federation must bind the units "always to fight on the same side; and if they have such feelings towards one another, or such diversity of feelings towards their neighbours, that they would generally prefer to fight on opposite sides, the federal tie is neither likely to be of long duration, not (nor?) to be well observed while it subsists".

B.6. *Existence of groups of units following contrasting and mutually incompatible social, economic, and political philosophies.* Such was the division in the United States between the slaveowning and the so-called "free" States. Such was also the division between Southern Rhodesia and the two other units of the Central African Federation (Dikshit, 1971b).

Some General Hypotheses

I. If all the various factors for unity enumerated above in set 'A' were present, there would be no need to bother about federalism, for, as Jennings (1953, p.55) wrote: "nobody would have a federal constitution if he could avoid it". Such a group of communities should, and most likely would, evolve into a unitary State.

II. Conversely, if all the factors for separation enumerated above in set 'B' were present, there can be no ground for union. The communities involved should, and in all probability would, organize as separate sovereign States forming at best, some limited purpose alliance or association.

III. Only when some factors of set "A" and some of set "B" are jointly present, would there be the need for a federal union—that so-called halfway house between complete unity and complete separation.

To put these proposition in symbolic language, if

U = Unitary State;
N = No union;
F = Federal union;
F_c = Centralized federalism;
F_p = Peripheralized federalism;
A = Factors for unity; and
B = Factors for separation;

then

$$A_{1, 2, 3, 4, 5, 6...} = U$$
$$B_{1, 2, 3, 4, 5, 6...} = N$$
$$A_{1, 2, 3, 4, 5, 6...} + B_{1, 2, 3, 4, 5, 6,...} = F$$

The different factors of the sets "A" and "B" (*i.e.*, those for unity and separation) may be present in different combinations. Accordingly, therefore, the nature of the resultant federations would also differ. From a study of the federal partnerships experimented so far, some of these possible combinations and the nature of the resulting union can here be envisaged. On the basis of these, certain tentative hypotheses on the origins of federalism may be put forward:

1. If the first factor of the first set (*i.e.*, the factor Ai) is found in combination with the first three factors of the second set (*i.e.*, factors $B_{1,2,3}$), the situation is confederal. A union consummated under such a condition, would last only so long as factor A_1 (that is, external threat) is formidable and is so recognized by all the parties to the union. Pakistan offered a good example of this type. Such a situation can, nevertheless, change into a federal one if, the actual or presumed threat to the survival of the constituents remains in existence for a sufficiently long period of time to inculcate among them the habit of cooperative action. During this period, through wise statesmanship and planning, the parties involved may succeed in creating some kind of an interdependent economic arrangement mutually beneficial to all the constituents.
2. If factors 1, 3, 4, 5, and 6 of set "A" are present in combination with any one of the first group of set "B" (language, race or religion, *i.e.*, B_{1a}, B_{1b}, or B_{1c}) a fairly centralized and articulate type of federation is likely to result. India offers a good example.
3. If factors 1, 3, 4, 5, and, 6 of "A" (*i.e.*, $A_{1,3,4,5,}$ and 6) are present along with an identical distribution of diversities in respect of the three elements (*i.e.*, language, religion and "race") in the first group of set "B" (that is, $B_{1 (a+b+c)}$), the resultant union may still be a centralized federation,

having strong dualistic undertones. Canada is a good example. The *raison d'etre* of such a federation would depend upon factors 1 and 3 of the set "A". If at any stage in its evolution, either of these factors were to disappear, the sense of unity in the federation is likely to be adversely affected. If both these factors were to disappear together, the union would in all probability collapse.

4. If factors 1, 2, 4, 5 and 6 of set "A" are present along with factor 2 of set "B" (*i.e.*, the existence of units with independent or competitive economies), a federation may still be born. Although, its survival will largely depend on the continuance of factor A_1, *i.e.*, the presence of a common external threat to collective security. If external threat is not present for a long enough time to have forced the constituent units of the federation to become adjusted to the new situation to their mutual economic benefit (that is, if they have not succeeded in creating interdependent economies during the period that the external threat has lasted), the end of this threat may mean the end of the federation itself. No historical example of this type is available, though the West Indian example reaches close to one. As will be recalled that the West Indian federation was originally initiated by indigenous leaders in order to achieve independence from British colonial rule. When it became clear that independence could be achieved even without a federation, the federalistic enthusiasm began to wane. Similarly, it is not difficult to appreciate the force of possible external threat as a factor in the two decades of continued union of the two wings of former Pakistan.

5. When all or most of the factors of set "A" are present, along with factor 3 of the set "B" (that is, if units concerned have regional identities based simply on their prior existence as autonomous provinces or as units of an overall system), in other words, if the units do not have any significant ethnic or economic bases for their regional identities, the resultant union may be a centralized federation with a high degree of federal

articulation. This situation shall prevail provided, after the inauguration of the federation, political and military conditions in and around the State remain largely peaceful, so that the units are not required to forgo their regional autonomy for the sake of national security. Granted peaceful conditions during which the units were left to mould their own local/regional administration, the units tend to develop separate identities and vested interests around them. And once they get into the habit of being the masters of their own houses, they are reluctant to subordinate themselves completely to the central government. West Germany offers a good example.

6. If two, or all of the elements of B_1 are present along with most of the factors in set "A" (except the factor A_2) and the patterns of diversities created by B_{1a}, B_{1b}, and B_{1c}, (*i.e.*, language, religion and "race") overlap in their regional distribution in relation to the boundaries of the constituent political units, the resultant union is likely to be a centralized federation. It will be so because the overlapping patterns of the different diversities tend to divide each conflicting group against itself, since, despite the antagonism that a group may harbour against a rival group of another type, neither group is able to overplay its hand since sizeable sections of its own group may possess strong sympathies with certain sections in the opposite camp. Such a pattern of distribution of diversities, by its very nature, forces the relevant communities to compromise and to learn to live and let live. Switzerland offers the best example.

7. The presence of factor B_5 in whatever combination of factors in the sets "A" and "B" will not produce a stable union. If the units involved recognize conflicting threats to their safety and security, no durable union is likely to emerge, whatever other factors might be present. The Central African federation provides a convincing example.

8. The sixth factor in set "B" (*i.e.*, the existence of groups of units following contrasting or mutually incompatible

social, economic and political philosophies), in whatever combination of factors in set "A", creates an essentially confederal situation. A federation created under these conditions can last only as long as the constituents remain convinced that they derive greater economic and other dividends within than without their union. The factor B_6 will in the long run, either break the union or would itself have to be broker-(by coercion) so that the union could be maintained. The United States of the Civil War days would prove the point.

To summarize the points stated above, if

U = Unitary State;

N = No union;

F = Federal Union;

F_c = Centralized federalism;

Fp = Peripheralized federalism;

A = the factors for unity; and

B = the factors for separation;

then the general propositions are:

$A_{1,2,3,4,5,6,...}$ = U (union)

$B_{1,2,3,4,5,6,...}$ = N, (No union)

$A_{1,2,3,4,5,6,...} + B_{1,2,3,4,5,6,...}$ = F (Federal union)

Based on various combinations of factors in sets "A" and "B", the following models on the origins of federalism may be deduced:

(1) $A_1 + B_{1,2,3,...} = F_p$ (Peripheralized federation)

(2) $A_{1,3,4,5,6} + B_{1a/1b/1c} = F_c$ (Centralized federation)

(3) $A_{1,3,4,5,6} + B_{1(a+b+c)} = F_{cp}$ (a centralized federalism with strong dualistic undertones)

(4) $A_{1,2,3,4,5,6} + B_2 = F_p$ but $A_{2,4,5,6,} + B_2 = N$

(5) $A_{2,3,4,5,6} + B_3 = F_c$ but $A_{1,2,3,4,5,6} + B_3 = F_c \rightarrow U$

(6) $A_{1,3,4,5,6}\ B_{1a} + B_{1b} = F_c$

(7) $A_{1,2,3,4,5,6} + B_{5...} = N$

(8) $A_{1,2,3,4,5,6} + B_{6...} = F_p \rightarrow N$

A Final Point

A scrutiny of the foregoing propositions would reveal that in all the combinations of the factors for unity and separation envisaged above, a necessary condition for the birth of a centralized federalism is that at least one of the first three factors in set "B" must be present. That is, in order that the resulting centralized union is federal rather than unitary, it is necessary that the political units concerned possess some regionally-grouped diversity as the *raison d'etre* for regional autonomy. Such regionally-grouped diversities may result either from regional differences in the ethnic composition, or from the existence of competing economies in the constituent units. It may also sometimes result simply from the fact of prior existence of the units as separate or autonomous organic political entities. The precise nature of the resultant union would depend upon which of these regionally-grouped diversities forms the basis of the regional identities. The critical factor to remember is that, only when the communities that are uniting together, possess separate regional identities, based on some regionally concentrated diversities, would the resultant union, under normal conditions, be federal. Indeed, federalism would often become nothing if it is required mainly to cope with diversities that lack inclusive territorial bases. A federal union requires that the communities involved should, besides recognizing a common national boundary (political, cultural and other) should also recognize some well defined cultural, economic or any other significant boundaries along the lines that separate them from one another, but the regional political boundaries need not strictly coincide with the regional distribution of cultural, ethnic or other diversities within the State. Thus, not diversity per se, but regionally-grouped diversity within unity is the essence of centralized federalism.

GEOGRAPHICAL FACTORS IN THE FEDERAL SYSTEM

"A federation", wrote K.W. Robinson (1961), "is the most geographically expressive of all political systems. It is based on the existence of regional differences and recognizes claims of the component areas to perpetuate their individual characters.... Federation does not create unity out of diversity, rather it enables the two to coexist." All political systems have a geographical basis but it is most prominent in the federal system as according to Robinson a federation is the most geographically expressive of the political systems.

The geographical factors that seem to have favoured the establishment of federal states, are :

(1) vast area,

(2) spatially grouped diverse population,

(3) odd shape and

(4) presence of several core areas.

A federation is based on the presence of a number of regionally grouped communities within the state's geographical area, which have some common interest to form the Union. .At the, same time these communities, possessing "regional strength and pride" have their own separate interests and desire to have the freedom to safeguard those interests. Although, there can be no direct relationship between the presence of regional diversities and the size of the state, it is most likely that a large state possesses greater regional diversity, which is a fundamental element in the federal system.

The vastness of area is one important geographical factor in the federal system. "All large states except China", wrote Pounds (1963), "have a federal constitution at the present time". However, he added that "several small countries, in which there is strong regional feeling have also adopted the practice of having a federation".

The geographical factor of size has thus a strong influence on the form of government. K.W. Robinson (1961) wrote that "countries, of large area and small population or even of rather large population, concentrated in widely scattered areas, are obviously suitable for the federal system". According to de Blij (1967) "theoretically the federal framework is specially suitable for states in the large and very large size categories. Poor communication and ineffective occupation of large areas within the state still affect, for example Brazil and Congo (Zaire). Such impediments to contact and control might disrupt A unitary state, while a federal framework can withstand their, centrifugal forces." In large and diverse states such as U.S.A. and Canada the danger of centralization of political power is real. Although, great development of transport and communication has markedly shrunk both countries, especially the United States they are distinctly divided into several regions. There is no alternative to the federal system in these large states. The observation of de Blij that large and comparatively sparsely populated states of the world are federal states is not without meaning though there may be a few exceptions to this rule. If large size favoured the rise of federalism in the world, then size and federalism have certainly a cause and effect relationship.

As compared to size, the shape of the state is insignificant in its political geography. One will, therefore, ask whether shape of political unit can be a factor in the origin of federal system, de Blij has held certain shapes responsible for the federal system of some states. "In terms of shape", writes de Blij, "it is obvious that fragmented states and states with pronounced and important proruptions may be best served by a federal system.

An elongated state possessing more than one core area also sought to turn to a-federal arrangement". In several cases the shape has influenced the functioning of the state. The Malaysian federation, which has a fragmented shape is a case in point. Malaya on the mainland and Sabah and Sarawak on the Borneo Island are separated by thousand miles of sea. These states, therefore, decided to form a federal union. It may, however, be remembered that the shape was not the only factor in the establishment of the Malaysian federation.

Large size as such may not cause a federation, as in the case of China. But it does work through giving rise to spatially grouped diversities. The larger a state is,, the greater is internal diversity. Different peoples who do not want to be under the unitary government get federally united for some common purpose. The 'original differences arise from the multiplicity of race, religion and language. But these aspects of diversities must be spatially grouped so as to cause a federation. It is in those countries where ethnic, linguistic, religious or cultural differences of the people have regional expression that various people see individual parts of their country as a homeland. For these countries federal arrangement has offered a political solution to the problem. The former Yugoslavia had, for example, several ethnic groups—Slovenes, Croats, Serbs and Muslims, who formed the federation. The former U.S.S.R., a mufti-ethnic, multi-religious and multi-lingual state, became a federation after the world war I. In these states of strong regionalism, federalism offered a governmental device, in which state autonomies went a long way to weaken the centrifugal forces and to strengthen the centripetal Ones in holding the state together.

The multicore nature of the political area has given rise to federalism in several parts of the world. "Federal states can adjust to the presence of more than one core area (or a number of subsidiary areas) more easily than unitary states". Several federal states of the world are multicore states, de Blij has cited examples of Australia, Canada, Nigeria and pre-1971 Pakistan to stress the point that only the federal constitution of these states could hold the diverse regions together. The multi-core federal states are characterized by the presence of a number of population clusters, separated by sparsely populated and unproductive areas.

TYPES OF FEDERAL STATES

Harm de Blij (1967), on the basis of the origins of federal states, divided them into five types :

(1) mutual interest,

(2) compromise,

(3) centralized,

(4) imposed, and

(5) shape (induced).

He has been criticized for using an arbitrary division of federal states. However, it throws some light on the nature of federal states.

In some federal states, population is homogenous as it is characterized by a common language and ethnicity, while in others people differ in some important respects. In each case people: faced the same danger and their mutual interest lay in facing the danger jointly. Switzerland, Australia, U.S.A.,

Canada and Brazil are mutual interest federations. The Canadian federation, being bi-lingual, required certain degree of compromise in its establishment.

Some federal states were established as a result of the willingness of the participating people to safeguard the interest of a large state at the cost of the individual interests. The Indian Union required a great amount of compromise and adjustment on the part of multiethnic and multilingual people. The Union of India was formed with a democratic framework and a well thought federal constitution. Nigeria is another compromise federation.

The centralized federal states are more unitary than federal. They function as centralized states inspite of their federal constitution. The former U.S.S.R. and former Yugoslavia are examples of centralized federal states. The over-centralization of forces in the federal government was possibly one of the factors for their collapse in 1992.

There are a few federal states on which federal system was imposed much against the wishes of the majority people who were involved. They were established either by a powerful minority or a colonial power. As a consequence of imposition such federations were short-lived. The Mali Federation, the Central African Federation and West Indian Federation rose and fell. The Malaysian Federation and Federation of Ethiopia and Eritrea are present day examples of the imposed federal states.

In some cases the shape—fragmented or prompt—forced; the political leaders to opt for the federal polity for the solution of the problem arising out of an odd shape. At the time of the birth of Pakistan in 1947, the country was fragmented into two major units separated by some thousand miles of Indian territory. The new state of Pakistan had to adopt the federal form of politico-territorial organization for forging unity of Western and Eastern wings. "It is doubtful" write de Blij that Pakistan would have become a federal state, were it not fragmented into two major units". It is noteworthy that with the exception of Pakistan and Malaysia, Islamic states of the world are unitary and highly centralized. Congo (Zaire) has a prompt shape and possesses several widely separated core areas located in its vast interior. Only a federal set up could hold the country together. If Malaysia is to be a unified state, then only a federal arrangement can possibly solve the problem of its scattered peninsular-insular territory.

It may be noted that the above types are not mutually exclusive, since two or more elements of federalism may be recognized in some federal states. For example, the Malaysian Federation is both imposed and shape-induced. Likewise Canada is a mutual interest and compromise federal state.

FEDERAL STATES OF THE WORLD

Norman J.G. Pounds provided list of 16 federal states that existed in the early sixties. They were 11.3% of the total number of 141 states. Their continent-wise break-up is given below:

Asia	3 India, Pakistan, Malaysia
Europe	4 Switzerland, U.S.S.R., Yugoslavia, West Germany
North America	3 U.S.A., Canada, Mexico
South America	2 Argentina, Brazil
Africa	3 Nigeria, Ethiopia, Zaire (Congo)
Australia	1 Australia.

The federation of Ethiopia and Eritrea which was formed in 1952 had ended in 1962 when Eritrea was fully integrated with Ethiopia. After two decades in 1992 two federal states—the U.S.S.R. and Yugoslavia—disintegrated giving place to a number of independent states. As such, today there are only 13 federal states which form 6.8% of 191 states of the world.

The federal system is a new Phenomenon in political geography. The modern federal practice started in U.S.A. in the 18th century; when American colonies had revolted against Great Britain. The thirteen states "with common language, mutual culture; deep local vested interest and a distrust of strong centralized government" declared their independence from Great Britain and formed the United States of America on July 4, 1775. The U.S.A. now comprises of 50 states and is the largest federal state of the world after disintegration of erstwhile U.S.S.R.

Canada is a bilingual federal state. Apart from the large area, the basic problem in Canada was the existence of English speaking communities of Ontario and Maritime Provinces, separated by Quebec, inhabited by French speaking people. "The unitary system of government that was set up to force unity upon a reluctant Canada did not prove to be successful, and in 1867 the federal constitution was adopted which, with modifications, has lasted until today" (Pounds, 1963, p. 34).

In Australia the British colonies of Queensland, New South Wales, Victoria, Tasmania, South Australia and Western Australia with population predominantly English speaking and with English culture—together formed in 1901 a federation (The Commonwealth of Australia) with the common purpose of facing the danger of imperialism from the Pacific. The Australian Federation has one federal territory named as Northern Territory, formed from South Australia in 1911 and one Capital Territory, known as Australian Capital Territory (A.C.T.), in which Canberra the capital city is located. These two have existed as separate political units since the dates of their creation.

The erstwhile U.S.S.R. was formed by the union of Russia, Ukraine, Byelorussia and the Trans Caucasian Soviet Socialist Republic in 1922. The process of formation continued upto .1940. In 1925 Uzbekistan and Turkmenistan and in 1929 Tadzhikistan became union republics of the U.S.S.R. In 1936 a new constitution was adopted. Transcaucasian republic was split up into three republics of Armenia, Azerbaijan and Georgia. In the same year Kazakhstan and Kirghizia autonomous republics of the Russian Federation were made union 'republics'. In August 1940 Moldavia and three Baltic States of Estonia, Latvia and Lithuania joined the Soviet Union as Union republics raising the number of republics to 15. This largest federal state of the world existed for more than five decades. It disintegrated in 1992 and in its place the Commonwealth of Independent States (C.I.S.) came into existence.

Three states of Latin America (Argentina, Brazil and Mexico) have federal constitutions. Argentina became an independent country on gaining freedom from Spain in 1816. The first constitution was passed in 1853 and a stable government was established. The Argentine Republic consists of 22 provinces, one federal district and the national territories of Tiera del Fuego, the Antarctic and South Atlantic Island. "The provinces erect their own legislatures and have general control over such affairs as are not under the jurisdiction of the Central government." The federal capital, Buenos Aires, covers an area of 200 sq. kms.

The federal constitution of Brazil, adopted in 1946, named the country as the United States of Brazil. Under a new constitution which came into force in 1969 the name was changed to Brazil and later to Federal Republic of Brazil (Republica Federativa de Brazil). The large and internally diverse country consists of 22 stales, 4 federal territories and 1 federal district and new capital Brasilia lies in the recently demarcated federal district. "Each state has its distinct administrative, legislative and judicial authority, its own constitution and laws which must however agree with the constitutional principles of the Union. The states may unite or split or form new states".

Mexico is a federal republic. A new constitution, amending toe old one of 1857, came into force in 1917. It has been amended from time to time. The republic is divided into 31 states and 1 federal district, each of which has the right to manage its own local affairs.

In Africa, federal system of governance was tried to serve mutual interest in several cases. It, however, did not work in the Mali Federation and the Central African Federation of Rhodesias and Nyasaland. Both were, therefore, very short-lived. The federation of Ethiopia and Eritrea, was brought into existence on September 15, 1952 when Eritrea, a former Italian colony (under British military administration since 1941) was handed over to Ethiopia in accordance with the resolution of the General Assembly of the United Nations Organization dated December 2,1950. Eritrea became an autonomous unit within the Federation. Ethiopia became a unitary state on November 14, 1962 when Eritrea was fully integrated with Ethiopia. (Eritrea recently became the 52nd state of Africa when it formally seceded from Ethiopia. TOI, Lucknow, June 24,1993).

Nigeria became a federation under a Governor-General on October 1, 1954 even when it was under the British colonial rule. It became a sovereign and independent country in 1960 and three years later was declared a Federal Republic. Nigeria is the most recent federation. If has four federated regions—North, West, Mid-West and East—which are based upon tribal groups and adjusted' to the ethnic and cultural diversities of the country.' The federal capital is located at Lagos.

The Congo (Zaire) emerged as a loosely federate state on gaining freedom from Belgium in 1960. Like those of Nigeria the federal units coincide with tribal groups—"Negroes, Pygmies and Hamites'. Because of areater size and diversity than in Nigeria, the Congo republic is rather gloose confederation. There was no alternative federalism because of vast size and prorupt shape. "The constitution promulgated in 1962 reserves to 21 provinces those powers which it does not explicitly reserve to the centre or does not place upon a concurrent list to be dealt with by both the centre and provinces".

Three small European countries have federal constitutions. The Swiss Confederation is the oldest in the world. It started in 1291 with a loose alliance of four cantons. The basis of the confederation was the profit derived by the federating cantons from the brisk trans-Alpine trade. Later on other cantons joined the confederation which attained its present limits by the mid-16th century. Originally Switzerland was "the loosest possible form of federation". But it gradually acquired a tighter organization. In 1848 the country adopted a federal constitution in place of earlier confederation. Since then the Swiss federation has remained stable and there is no possibility of its collapse. "Switzerland is a federal state of great internal diversity (with four languages—German, French, Italian and Romansh) and yet its stability is unequalled. Alone in Europe it retains virtually unchanged boundaries of a century and half ago. Alone in Europe it has been neutral through decades of violent conflict."

"In Germany as in Switzerland the federal structure is explained by historical tradition rather than geographical diversity". The Federal Republic of Germany had evolved from a loose agglomeration of states—the German Confederation of the 19th century. The federal organisation of government had survived through a long period except for a brief spell when Hitler imposed a unitary system of government After its defeat in the Second World War, Germany was divided into West Germany and East Germany. In 1949 West Germany regained its federal constitution and in 1955 its sovereignty from the Allied Powers (U.S.A., U.K. and France). In 1990 the German Federal Republic (West .Germany) and German Democratic Republic (East Germany) were united together and the federal organization was extended to the eastern parts with Bonn as federal capital.

The Yugoslav state was established after the World War I in 1919 as a unitary state. It had a variety of related peoples, Serbs, Croats, Slovenes, and Muslims, together with smaller communities. The first constitution of 1919 failed unite the diverse ethnic groups. It was, therefore, replaced by a federal constitution in 1939 which gave regional autonomy to the major ethnic groups.

.The Constitution of 1953 created six republics of Serbia, Slovenia, Croatia, Bosnia-Herzegovina, Montenegro and Macedonia, and two autonomous areas. The constitution was amended twice, first in 1963 and again in 1974. The federal system operated in a centralized manner under the leadership of Marshal Tito. The constitution provided autonomy for the six republics but "failed to define clearly the fields in which this autonomy might operate." Federalism therefore could not survive after the death of Marshal Tito. The Yugoslav federation broke up in 1992 to give birth to five independent states.

The three federal states of Asia arc Malaysia, Pakistan and India. The federation of Malaysia came into being on September 16, 1963 when the Federation of Malaya was broadened to include the state of Singapore, the British possession of North Borneo (renamed Sabah) and Sarawak. The U.K. relinquished sovereignty over Singapore, Sabah and Sarawak in July 1965. Later in August 1965 by a mutual agreement between Malaysia and Singapore the latter seceded from the Malaysian Federation. The federal capital is Kuala Lumpur, established on February 1, 1974 with an area of approximately 94 sq. miles.

The Malaysian Federation is the best example of a federation imposed by a colonial power and fragmentation of its peninsular-insular territory into several major units. "Any glance at the area of Malaysian Federation must produce fears of its future"*

The state of Pakistan came into existence as a result of partition of India in 1947 and consisted of two major units, separated by foreign land. Pakistan became a federal republic in 1970 when the Legal Framework Order was published denning a new constitution, with a Moslem Head of State, the National Assembly and provincial assemblies. In December 1971 eastern province (East Bengal) declared itself an independent state of Bangladesh. The new constitution of Pakistan, enforced on 14 August 1973, provided for a federal parliamentary system with the President as Constitutional Head and Prime Minister as Chief Executive. The Federal Republic of Pakistan now consists of North-West Frontier Province, Punjab, Sindh, Baluchistan, Federally administered Tribal Areas and Federal Capital Territory of Islamabad. It may be noted that even after the secession of East Bengal (Bangladesh) Pakistan opted for a federal constitution. This reduces the significance of shape factor in the origin of federal states to a considerable extent.

THE BREAK-UP OF MULTI-ETHNIC FEDERAL STATES

The early 1990's witnessed the break up of two federal states, namely ' U.S.S.R. and Yugoslavia. The demise of the Soviet Union, the world's biggest federal state comprised of 15 union republics, is however the most significant development of the early nineties. In August 1991 in a failed coup some Russian leaders attempted to overthrow President Mikhail Gorbachov. The president of Russian Federation, Boris Yeltsin first came to his help but later withdrew his support in preventing the break up of the Soviet Union. In September 1991, three Baltic states—Estonia, Latvia and Lithuania—seceded from the U.S.S.R. and as independent states, were admitted to the United Nations Organization Three months later in December 1991, the Soviet Union finally disintegrated and gave birth to the Commonwealth of Independent States (C.I.S.), which is comprised of 10 states, namely, Russia, Ukraine, Belorus, Kazakhstan, Kyrgystan, Tajikstan, Turkmenstan, Uzbekstan, Moldova and Armenia. Two republics—Georgia and Azerbaijan—did not join the C.I.S.

The factors which possibly led to the disintegration of the Soviet Union were the advanced State of decay of the Soviet economy, the static form of Soviet political system, the spread of virulent form of nationalism and mounting pressure of the West, specially the U.S.A.

The former Soviet Union was a conglomeration of nations, kept together by the dictatorship. It was a super state, never a nation state. Its disintegration demonstrates the fact that states break up "because of economic, or political non-viability or because of economic or political colonization." "The former Soviet Union was, economically unviable, unable to produce, export and prosper except in the field of armaments..... The Soviet political system was (also) becoming anachronistic on the threshold of the 21st century."

Yugoslavia is another example of a multiethnic state, which has disintegrated into five independent states. Among the several ethnic groups Serbs, Croats, Slovenes and Moslems formed more than 80% of the population of former Yugoslavia. Under the federal set up of 1953, the Federal Peoples Republic of Yugoslavia was comprised of Six republics,' namely, Serbia, Montenegro, Slovenia, Croatia, Bosnia-Herzegovina and Macedonia. In July 1991 Slovenia and Croatia broke away from Yugoslavia and these two along with Bosnia-Herzegovina were admitted to the U.N.O. in May 1992. Macedonia has not yet been able to secure membership of the U.N.O., because of the dispute over its name with Greece. Montenegro chose to remain with Serbia and Serbia-Montenegro now represents the rump state of Yugoslavia with Belgrade as the capital city.

Evidently there was no such thing as a Yugoslav nationalism. Moreover, a geographical merger of various ethnic groups could not develop into one nation. The Croats, the Serbs and the Bosnian Muslims agreed to coexist in Marshal Tito's Yugoslav federation. After his death there was no leader of the stature of Tito, who could keep the diverse peoples united. The break up of the former Yugoslavia in 1991 was undoubtedly the product of a leadership crisis.

THE GEOGRAPHICAL BASIS OF INDIAN FEDERALISM

"In India," writes de Blij, "the federal idea emerged as the country moved towards self-determination (rule) and ever since independence the state's federal structure has been changed (by amendments of the Constitution) to lessen the problems of the country's great internal diversity". As a matter of fact vast diversities which are spatially grouped within the territorial limits of the country form the geographical basis of Indian federalism.

Large Size and Very Large Population

India is a country of great dimensions. She is the second largest country in population (840 million) and seventh largest in area (3.27 m sq. km.) in the world. The vast size in an important factor in the adoption and continuance of the Indian federal system. Only a" federal constitution can better serve the interest of the people of different parts of the country.

Great Diversity

India is a land of great linguistic and cultural diversity. Its inhabitants follow four major religions and speak about two dozen major languages. Several linguistic regions came into existence as a result of multi-lingual character of population. Language acted as a great centrifugal force and only the federation could hold the constituent units (linguistic states) together.

The country is beset with problems of religious diversity. It was partitioned on religious lines in 1947 as a result of mixing of religion with politics. Religious diversity was however only a minor factor in the federal organization of India. The federal constitution has provided ample safeguards to the religious minorities.

Ethnic diversity may have been the basis of federalism elsewhere in the world. The former Soviet Union and Yugoslavia became federal states and disintegrated due to multi-ethnicity. But in the case of India ethnic complexity was not at all the basis of federal system. Only in the north-east different ethnic (tribal) groups clamoured for regional autonomy. The formation of Nagaland and Maghalaya illustrates the case in point Thus ethnicity influenced the federal polity to some extent.

Complex Physical Framework

India has three distinct major physical divisions—mountainous region of the extreme north, great plains of the north and southern plateau region. Then there are empty mountainous and desert areas. At the same time great plains of the north support one of the great human agglomerations of the world. Due to physical barriers and great diversity of relief, a large number of macro and micro physico-geographic regions developed in isolation. Political fragmentation and regionalization were superimposed and entrenched on the physico-cultural landscape even prior to the British, colonial rule. 'Before independence India was a centralized; and unitary state and provinces in which the country was divided, were established on the basis of administrative convenience.

The architects of post-independence India were therefore faced with a difficult situation. Their task was to keep all regions of the country united into a nation-state. They completed this task by providing a federal constitution and linguistic organization of states. "From an early stage it was clear that an independent India could survive only as a federal state— a union of diverse territorial units."

The Indian leaders decided to give the country a constitution which could satisfy different people and to bind diverse regions into a working federation. At the time of independence India's territory was divided into 12 provinces and 562 princely states. The merger of princely states into the Union of India either as groups of states or parts of the province was completed by the end of 1949. "A federation perforated by hundreds of non-participant entities clearly was not a viable proposition". Their integration into the Indian Union was, therefore, necessary for the unity and integrity of the country.

NATURE OF INDIAN FEDERATION

The Indian Constitution mentions that India shall be a 'Union of States'. The use of the term 'union' instead of 'federation' has raised questions whether India is a federal state or not. Several political and legal experts have described the Indian Constitution as federal while some others opine that it is less federal and more unitary. However, it is in reality a quasi-federal constitution. The farmers of the Constitution therefore avoided the use of the term federation. Indian federation is not pure federation in the strict sense of the term.

The Constitution of India has, however, several federal features. Firstly, it has provided for a dual polity by establishing two types of governments—the Union (Central) government, and State governments. Secondly, it makes a division of power between the centre and the states. As already mentioned above, there are three lists—Union list, State list and Concurrent list of subjects. The Centre and States have powers to be exercised in the fields assigned to them by the Constitution. Thirdly, there is a provision for an independent judiciary for interpreting the Constitution. It would thus be seen that the Indian Constitution is basically federal. It has, however, some unitary features and strong bias towards the centre.

The states of India have a subordinate position in the federal system. It is perplexing to note that they are also destructible territorially, whereas in Switzerland, U.S.A. and Australia they are not so. It is perhaps, due to the fact that the Indian federation did come into existence as a union of independent states as in the case of the U.S.A. or Australia. Moreover, most the states were linguistically heterogenous and the Indian National Congress, Which spearheaded the freedom movement, favoured the reorganization of states on linguistic basis.

The changes in states were necessary because the framers of the Constitution realized that the process of change could not be postponed for long. As such it was not necessary to continue with the initial organization of states (categorized as Part A, B, C and D) as mentioned in the Constitution when the people demanded a change. Hence, the Constitution provided for the formation of new states and alteration of areas, boundaries and names of existing states. The parliament may by law form a new state by separation of territory from any state or by uniting two or more states or parts of states or by uniting any territory to a part of any state.

It clearly means that the area of a state may be increased or decreased and names and boundaries may be altered. As a result of this provision in the Constitution, the Indian Parliament has affected changes in the administrative map of India several times, first in 1956 when states were reorganized on linguistic basis into 14 states and 6 Union Territories. The Indian federation thus differs from the traditional federal system since it empowers the Parliament to alter the territory or integrity of states without their consent or concurrence.

As regards the status of states, they are no more than administrative units. Although, they derive legislative powers from the Constitution, the division of powers is such that the Union has a very tight control over them. The Constitution provides that toe state will legislate on all matters mentioned in the 'state list'. But in reality it is not so. The Parliament interferes with the legislative powers of the state" through relevant Articles of the Constitution. During national emergency Parliament can legislate on any matter of the state list. This means that the entire state list becomes a concurrent list. Even in respect of the concurrent list the law passed by Parliament prevails over the law passed by the state legislature. "It is a deviation from the basic principle of federalism, because in a federal, system the federal legislature (Parliament) cannot unilaterally transfer powers reserved exclusively for the constituent units.

Although, the Sarkaria Commission has recognized the supremacy of Parliament over the state legislature, it has recommended some amendments in the Constitution to give more power to states to enable them to amend Parliamentary laws on subjects of the 'state list' with the assent of the President of India. The Commission has also advocated the stregthening of the state's say in the concurrent list and the loosening of centre's hold on the Union list. The centre should consult the state before legislating on subjects of the concurrent list.

The states in India have a subordinate position not only in the legislative field but also in the administrative field. The Union government has the executive power provided by the Constitution to issue necessary directives to the state governments for securing compliance with the laws passed by the Parliament. If directions are ignored or resisted, the President can put the state under President's rule. The Central government keeps a close watch on the state administration through the officers appointed through die All-India Services. It can even appoint an enquiry committee against the erring Chief Minister. Moreover, it is through the Governor that the

Central government has a control over the state administration. He is appointed by the President on the recommendation of the Union government and holds office during the pleasure of the President. A state government can be dismissed by the President under Article 3.56 of the Constitution on a report from the Governor to the effect that constitutional machinery has failed in the state.

In a federal system the states must have sufficient resources to run their own administration. The Union government controls most of the financial resources of the country and this has made the states financially weak. The distribution of finances is done by the Finance Commission appointed by the President, and the Planning Commission prepares the Five Year Plans of the states and allocates funds for their implementation. No state can afford to ignore the directives of the Planning Commission. This has further increased the dependence of the states, on the centre. "The Planning Commission has upset the constitutional balance of the federation by encroaching upon the state economy". In April 1993 Nagaland urged the centre not to curtail the financial requirements of the state in the name of austerity as it would adversely affect socio-economic development, of the state. The law and order, a state subject, could not be maintained properly as the functioning of the police force was hampered due to resource constraints. This clearly shows that states depend too much on central assistance in the matter of finances.

THE FUTURE OF INDIAN FEDERALISM

It has been rightly observed that "federalism is more a functional arrangement for cooperative action". Currently it is the cooperative federalism which makes the federal state strong, stable and prosperous. The centre and states must cooperate with each other in all fields and there should be cordial relationship between the centre and states and also between one state and the other state.

During 1947-1964, appropriately called Nehru era, the Indian National Congress was in power both at the centre and in the states. The centre-state relations were, therefore, largely cordial. However, there was some resentment against the centralization of power. But the state governments were pacified anyhow. It was after 1967 when non-Congress governments were formed in some states that the demand for a review of centre-state relations was made. Since then there has been a persistent demand for changes in the provisions of the Indian Constitution.

India has a federal polity constitutionally speaking but in reality the system has largely been unitary. Indian federalism remains weighted towards the federal (central) government. The Constitution of India itself has certain provisions which assign the states a subsidiary position. Earlier in this chapter much has been said on the legislative dominance, administrative subordination and financial starvation of states. This does not augur well for the future of Indian federalism.

There are several Articles in the federal Constitution which are the cause of strained relations and confrontation between the centre and the states. They seriously threaten the federal polity of the country. Special mention will be made here of the mode of appointment of Governors, dismissal of a state government by the President and provisions for financial emergency.

The Constitution (Article 155) gives the power to the central government to appoint Governors of its choice. Obviously such a Governor cannot act independently. He would protect the interest of the political party in power at the centre, much to the embarrassment of the opposition-ruled states. It is on the report of the Governor that the President issues proclamation for dismissal of the state government and for putting the state under President's rule (Article 356). The latest example is afforded by the dismissal of governments of the four B.J.P. ruled states of Uttar Pradesh, Madhya Pradesh, Himachal Pradesh and Rajasthan on December 15, 1992 as a sequel to the Ayodhya incident. Later on April 2, 1993, the M.P. High Court quashed the Presidential proclamation dismissing the M. P. government. The court observed that the proclamation was invalid and beyond the scope of Article 356 of the Constitution.

The Sarkaria Commission, set up in 1983 to suggest reforms for an equitable distribution of powers between the centre and states, has recommended that the Constitution should be so amended as to make it obligatory for the central government to consult the Chief Minister of the state before appointment of the Governor is made. The Governor should also enjoy constitutional protection against arbitrary removal by the President. The Commission has also recommended that Article 356 be used very sparingly as a last resort and President should refer the proclamation with the Union government's grounds to the Supreme Court for opinion as to its constitutional validity.

The states in India are heavily dependent upon the centre for their financial requirements. It is general feeling that the central government does not treat the state governments well in the matter of finances. "The centre has misappropriated many of the powers which ought to belong to the states in a federal set up. Today almost 85 per cent of the industrial sector is under the Union government and all the high revenue yielding consumer essentials are outside the ambit of the state sales tax. Even the Finance Commission deals with only 20 per cent of the nation's financial resources for devolution purposes".

As the finances received from the centre are inadequate for development and other purposes, the states clamour (or a greater share of the financial resources of the country. The Sarkaria Commission has therefore suggested an increase in, state's share in areas of revenue and greater control by states over industrial licensing,

Economic grievances of the people are perhaps the main cause of separatist movement that are going on in Bihar, Assam, hills of West Bengal and Uttar Pradesh and elsewhere in the country. Coincidentally, these movements are "ethnic-specific". The Gorkhas, the Bodos, the Jharkhandis and others have been agitating for autonomy since they desire to improve their economic lot in their own way. The regional political parties are growing strong by raising the question of economic grievances, resulting from an economic system heavily tilted towards the centre.

There is a need for amending certain provisions of the federal constitution for keeping the country united by reconciling the regional aspirations and identities with the unity and integrity of the nation. "In India the concept of federalism as a bond to withstand the forces of disunity is being (has been) put to severest of tests" (de Blij, 1991).

Some people, however, feel that the Indian federal system is on the verge of collapse. It may not be wholly true. But changes in the Constitution are needed for improving the centre-state relations. India has already started changing its economic framework drastically. Why not renovate the political structure (federal system) into something which works.

TOWARDS A THEORY OF FEDERAL STABILITY

Some General Tendencies

Stability or maintenance of federalism implies, in the first place, the opposite of the idea of secession, collapse, and disintegration. It also implies the absence of complete unification of government and politics in the State, that is, a type of unification that may obliterate the regional governments to the point where federalism is replaced by a unitary structure. The continuance of effective regional autonomy alongside increased centralization of the more important governmental powers and functions, are necessary and complementary factors for the survival of federalism. The one prevents the transformation of federalism into unitary government, and the other allows the central government to exploit the advantages of nationwide resource base so that the federal authority could fulfil its constitutional functions more effectively.

The following paragraphs summarize some general conclusions on the success or failure of federalism as a system, arrived at from a comparative review of the functioning of federal partnerships, both past and present. This general survey differs from several of its predecessors in two ways. Firstly, it is a more focused analysis of the problem of federal stability. Secondly, it takes a more comprehensive view of federalism as a form of human organization, involving a whole range of forces and factors—economic, social, political and cultural—that may contribute to shape its functioning, and influence the survival of the State as a federal union.

1. One fundamental factor in federal stability is that the regional identities on which the constituent units of a federation are based, should not, in so far as possible,

be mutually exclusive, even though they may be largely conflicting. When boundary lines of the regionally-identifying diversities overlap, the cleavages have a tendency to get transformed into linkages, and the result is a stable federal partnership. This point is most clearly brought out by the example of Switzerland. Overlapping boundaries of language and religion in that country have weakened both language and religion as divisive forces, since each linguistic group contains representatives of both the faiths, and of course, vice versa.

Hence, although the Catholics may harbour a distaste for the Protestants (and vice versa), they cannot discriminate or fight against them because each (the Catholic as well as the Protestant bloc) is divided within itself, in the sense that owing to the overlapping identities, sizeable numbers of one group may maintain close affinity with a sizeable section on the opposing side, owing to the ties of common language of communication, irrespective of difference in faith. Thus, neither side can fortify its camp without risking a fair segment of its own group in the opposite camp. Therefore, neither language nor religion has been able to overplay its hand and the fissiparous tendencies have cancelled out. (Fig. 4.1.)

When we compare Switzerland with Canada, the moral is easily drawn. Because both the French language and the Roman Catholic faith have been very highly identifiable with the province of Quebec in Canada, here the religious-linguistic division has been the source of continuing strain on the federal partnership. The point is obvious in the case of India also, where, ever since the creation of linguistic States in the 1950s, fissiparous tendencies have raised their ugly heads.

The point is also clearly made when we closely analyze the American partnership during the first 50 years of the nineteenth century. It will be remembered that the institution of slavery, divisive though it had always been, did not strain the federal tie to the breaking point, so long as social and economic cleavages in the Mississippi valley had cut across each other.

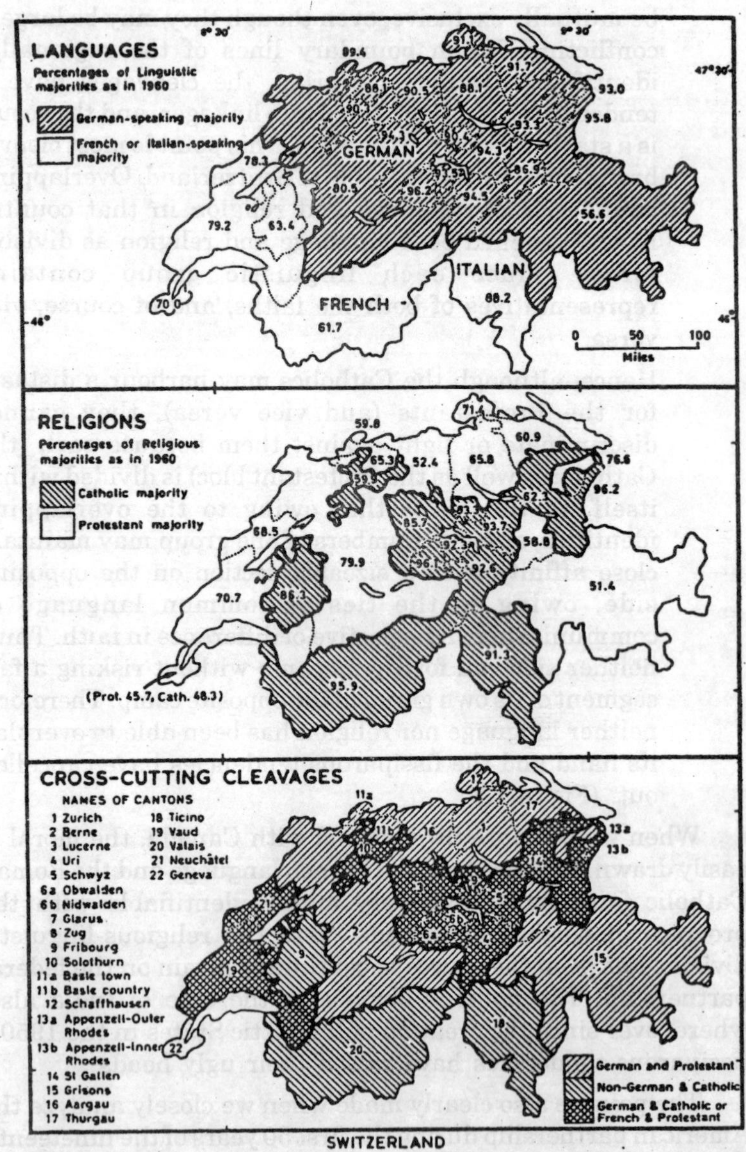

Fig. 4.1 : The overlapping cleavages of language and religion have neutralized each other's divisive force, and the cleavages have been transformed into linkages and, thus, have helped to create an urge for harmony and compromise. This has been the fundamental force in the process of States-formation and nation-building in this "multinational" society.

On the issue of slavery and freedom, most of the West was emotionally attached to the North. Economically, however, the entire Mississippi valley area (until about 1850) was very closely tied to the South through the arms of the great river, since it alone provided the main outlet for the commerce of this vast area.

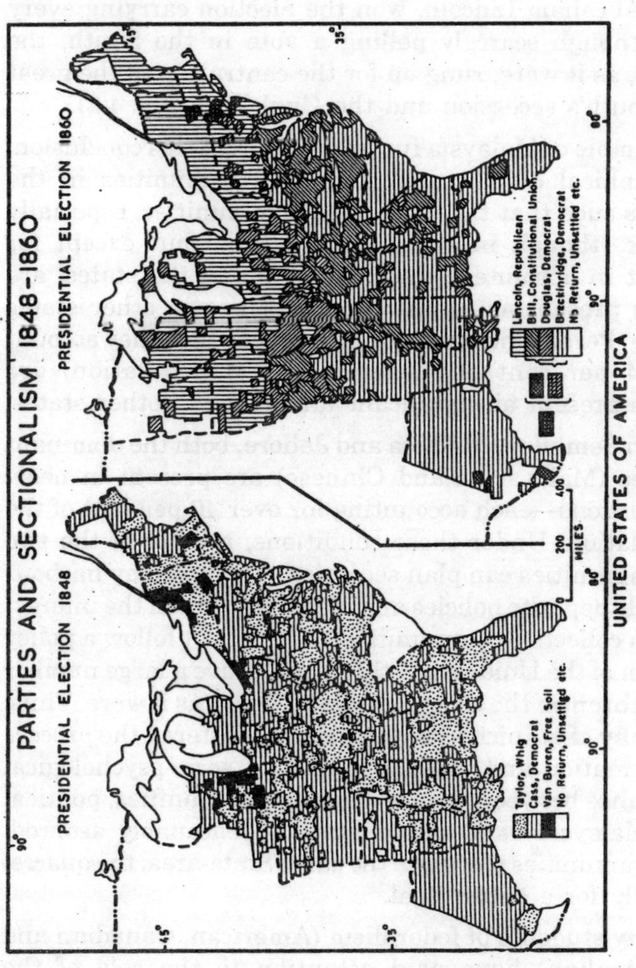

Fig. 4.2 : ***Parties and Sectionalism in the United States, 1848 and 1860:*** **Despite differences on the issue of slavery the nation had remained united so long as social and economic cleavages (until 1850) cut across the Mississippi valley Until then both the major political parties drew support from all parts of the country. But when in the 1850s railroads tied the upper West more closely to the North than to the South, the political cleavages became sharply defined almost along the Mason-Dixon Line, the line separating the slave-owing and free States.**

But in the next decade, as the railroads tied the upper West more closely to the North than to the South, the cleavage lines became sharply defined almost along the Mason-and-Dixon line—the boundary between the slave-owing and the free States. So sharp was the divide, that in the 1860 Presidential elections extension of slavery became the dominant issue. Political party divisions, which had so long cut across Regional boundaries now became closely identifiable with the sectional boundary between slavery and freedom. And as the new Republican candidate, Abraham Lincoln, won the election carrying every free State though scarcely polling a vote in the South, the curtain was, as it were, rung up for the central act of the great tragedy—South's secession and the Civil War (Fig. 4.2).

The example of Malaysia further reinforces this conclusion. The geographical distribution of racial communities in the federation is such that the immigrant communities, especially the Chinese, though in sizeable majorities (but except for Penang, not in absolute majority) in some of the states, are nonetheless present as significant minorities in other states (*i.e.*, Penang, Perak, and Selangor, in each of which they account for over 44 per cent of the state's total population) are nonetheless present as significant minorities in other states.

In Negri Sembilan, Malacca and Johore, both the dominant communities (Malaysians and Chinese) are present in near-equal proportions—each accounting for over 40 per cent of the State population. Under these conditions, neither of the two leading communities can plan secession. The two may harbour diametrically opposite policies and idealogies, but in the interest of their own collective communities, they cannot follow a policy of disruption of the Union since this would leave a large number of their brethren on the other side of the fence, as it were. Thus, the geography of ethnicity in Malaysia has fostered the process of State-formation and, despite a kind of socio-psychological uneasy balance between the two leading communities, political future of Malaysia as a united State is reasonably assured. Both the communities recognize the same State-area; the quarrel is essentially for a better deal.

2. Many students of federalism (American, Canadian and Australian) have paid attention to the role of the

structure of the political party system in the maintenance of centralization or peripheralization of federalism. Some of the best known studies on this topic are those by Truman and Grodzins on the United States, Wildavsky on Australia, and Muller on Canada. All these studies have focused attention on the role of the party system in the operation of the respective State as a federal system. William Riker, asserted (what was only implicit in the earlier studies) that

> the federal relationship is centralized according to the degree to which the panics organized to operate the central government, control the parties organized to operate the constituent governments. This amounts to the assertion that thc proximate cause of variations in the degree of centralization (or peripheralization) in the constitutional structure of a federalism is the variation in degree of party centralization. (Riker, 1964, p. 130).

> Whatever the general social conditions, if any, that sustain the federal bargain, there is one institutional condition that controls the nature of the bargain...This is the structure of the party system, which may be regarded as the main variable intervening between the background social 'conditions and the specific nature of the federal bargain (Riker, 1964, p. 136). (Emphasis mine.)

This line of approach has confused effect with the cause. Political parties or interest groups arise in response to certain inherent social and other related conditions obtaining in a given area—they cannot by themselves be an inherent part of the basic structure of any political system. The extent to which the party ruling at the centre would be in a position to control the governments in the regions would depend upon the nature of the geographical distribution of economic, social or cultural elements relevant to the political life of the nation. If such elements are characterized by somewhat exclusive patterns of regional distribution, the chances are slight that, under peaceful conditions, the party ruling at the centre would also control all or even most of the regional governments as well. But if the regional distribution of diversities on which the federal partnership is based is not exclusive, or the country has developed a strong nationally-conscious division of classes, the chances are greater that political parties would be nationwide organizations, so that the ruling party at the centre may also control most of the regional governments as well. Such a situation prevails in Australia where the "Australian Labor Party...in a sense...became a national party even before Australia was a nation". (Overacker 1952, p. 30).

The point was equally clear with reference to the Indian National Congress during its first flush of power under the leadership of the late Jawaharlal Nehru. The reasons were twofold: first, it was the party that had, in a sense, given birth to the new political nation. Secondly, it took some time for the new State units organized on a linguistic basis, to ossify as linguistic homelands of some sort. This ossification has now become clearly visible following the rise of linguistic patriotism. It is evidenced in the adoption of regional languages as the media of education up to the highest level, so that, the field of enterprise for the common man has got progressively restricted to his own linguistic homeland (*i.e.*, the state in case of states other than those that speak a language common with some other states, as is the case with the states in the Hindi belt). Hence, regional political parties have been able to capture power in a number of states. The number of such parties is likely to increase over time. Such a situation would not be easily conceivable in Australia or the United States outside the deep South.

Thus, the key factor in *federal stability is not the political party organization per se but the nature of the federal situation which conditions the nature and types of political interest groups.* If the social, economic and other cleavages cut across regional political boundaries, as in Australia and to a lesser extent in the United States, the political party organization is likely to be centralized. But if the cleavages in national life are somewhat watertight, as in Quebec vs rest of Canada, the result is bound to be just the reverse. This, however, is not to deny the role of the tangled feedback relationship between party action, party organization, and the complex of all the other factors which compose the "federal situation". But, while analyzing the role of party politics in the stability of federalism, it seems necessary repeatedly to refer back to the underlying socio-spatial forces that maintain political party structures in order that conditions may progressively be created to favour strengthening national parties vis-a-vis the regional ones.

3. In multi-ethnic federations with regionally grouped diversities, other things being favourable, the greater the factors of "felt" diversities in national life, the greater are the chances of a stable federalism. Under such a condition, chances are great that the spatial pattern of diversities would not coincide, and that their overlapping boundaries would neutralize their fissiparous potential.

   The point is well demonstrated by the example of India where despite a large number of apparent diversities, language has come to be recognized as the fundamental cleavage-forming factor at the level of the common man. As internal political boundaries have been made coincident with the linguistic ones, the cleavage lines between states have hardened, and language has begun to play divisive mischief. Chances for a more stable federal partnership would have increased if the country had possessed some cleavages conflicting with that of language. For instance, if, as some wrongly believe (Watts, 1966, p. 55), the universal system of caste among the Hindus had run across the nation uniting all those of similar caste together, instead of caste ties being limited by linguistic boundaries, (as in fact is the case), the divisive influence of language would have been very greatly neutralized.

4. In multi-ethnic or otherwise heterogeneous federations with strong regional identities, a continued external threat to the collective security of the nation is a favourable factor for centralized federalism. If external threat continues for a long enough time, the parties to the federal union, are compelled to adjust their differences and learn to live together as a close knit family. This factor had squeezed the discordant cantons of Switzerland into unity. The French, Spanish, and the

British threats had kept the otherwise discordant thirteen colonies together. The militant United States of the Civil War days had forced the two quarreling partners in "Canada" and the indifferent Maritime provinces to join together in order to create a united force against the encroaching Republic from the south. The point is well explained when we compare the two decades of unity in the old Pakistan with the rapid collapse of the West Indies federation. Conversely, opportunities for expansion may also foster federal sentiment by offering common rewards to all the constituents as was the case with the United States of America vis-a-vis Mexico, in the nineteenth century.

It has also been demonstrated that in areas far removed from active political frontiers, the sentiment for union is relatively weaker, as compared to areas bordering on the active international frontiers. Julius Pratt demonstrated it with reference to the United States of the 1812 War days (Pratt, 1957, pp. 126–127). The point is equally clear in relation to India. There is comparative indifference to the border problems with Pakistan (which do not seem to threaten the security of the nation as a whole, in view of Pakistan's relatively limited military power) in the southern states of the Indian Union, though when the nation was faced with the Chinese assault from across the Himalayas in the 1960s, the entire nation was alert to the danger.

In countries where federalism has been imposed over a largely nonfederal base (as was the case in the Latin American countries and in post-1945 West Germany), the situation is quite different. It is the lack of external military threat that would favour the continuance of the State as functioning federalism as against its transformation into a unitary organization. This is so because a federal political system once in motion, tends to create vested interests and states-rights sentiments, where none might have existed before. To enable new vested interests around the regional governments to take deep root in the popular consciousness, it is necessary that regional governments are left to manage their affairs without outside interference for a sufficiently long period of time. A comparison between the West German and Latin American examples would prove the point.

5. In most cases of failure of federalism (*e.g.*, Indonesia, the Central African federation, the West Indies, Pakistan, and Nigeria before the Civil War) one common factor in the breaking down of the federal tie was the presence of one unit numerically so large as to be capable of vying in strength with many or all of the others combined. Java was such a unit in the Indonesian federation, Southern Rhodesia in the Central African, Jamaica in the West Indian, East Bengal in the Pakistani, and the Northern region in the Nigerian federation before the Civil War. In fact, as Mill (1947, pp. 366-367) pointed out long ago, for the stability of a federal experiment it is necessary that "there should not be one State so much more powerful than the rest as to be capable of vying in+ strength with many (or all) of them combined. If there be such a one and only one, it will insist on being the master of the joint deliberations...", and thereby vitiate the federal partnership by turning the other components into its dependencies.

   The examples of East Bengal and Jamaica, however, show that numerical preponderance in population is not always a guarantee that the largest unit would predominate in the federal partnership. The overall locational advantages have much to contribute in deciding whether the largest unit would or would not become the dominant partner. And this would also largely decide whether the numerically dominant partner would or would not stay in the union, since, owing to its numerical preponderance, the numerically dominant unit is in a position to make its secession accepted.

6. Another point about federal stability is that there should be a sufficient number of federating units (Sawer [1969,

p. 132] thinks, probably at least five) to prevent continuous face to face conflict of one or two potential rival units. The point is best explained by the example of Nigeria before the Civil War, where the federal structure "with a three way only territorial dispersal of political authority, permitted or indeed encourage(d) two regions to combine and pit themselves against the third region in the struggle for the fruits of power" (Me Whinney, 1966, pp. 47). This strain on the federal relationship was further aggravated because of the numerical preponderance of one of the units—a situation which assured permanent membership of the political power bloc to the largest partner, *i.e.*, the North. The game of musical chairs generated by such a territorial pattern left the odd unit out of power at each turn. This brought the federation to a breaking point.

Where there are only two units in a federation, or only two clearly defined groups of units, there is likely to be a permanent tug-of-war between the units or the groups concerned. This situation is bound to put a great strain on the federation, and if the concerned units were to develop mutually incompatible regional identities, the federation would most likely break. The point was amply proved by the case of the United States during the Civil War, and the example of the two decades of uneasy partnership between the two wings of the united Pakistan.

7. In general, the greater the inter-regional mobility of population within the State, greater is the chance of federal stability. Secondly, for a continued federal partnership, it is necessary that the constituents continue to feel that they are economically better of within than without their union (Dikshit, 1975, p. 213). Federal stability requires that while loyalty to both the levels of government is retained, national loyalties must get precedence over the regional ones. A high degree of interregional mobility of population strengthens the sense of national identity by acting as an effective curb on divisive provincialism. This is perhaps the most

important lesson that the history of American federalism offers on this question. Her vast frontier of settlement made the United States right from the beginning a nation on the march. The continuous inter-regional movement of population among the states weakened regional loyalties and strengthened the national ones. Even when, towards the end of the nineteenth century, the frontier of settlement came to a close, inter-regional mobility of population continued unabated, since the physical "frontier" was soon replaced by the metropolitan one.

When we compare the examples of the United States with Canada, this hypothesis is easily confirmed. In Canada, the frontier could not act as an effective melting pot because French speakers in Quebec were generally reluctant to leave their province, owing to the lack of educational facilities in French. This lack of incentive for spatial mobility among the French-speaking population left the two cultural streams in that country to flow apart, so that despite well over a century of centralized federalism, Canada has not succeeded in creating a strong all-Canadian nationalism.

The absence of inter-regional mobility has been the greatest stumbling block in the rise of intense feeling for "nationalism" in most of the post-1945 federations. It was almost the undoing of the Nigerian federation. The absence of any scope for such mobility had all through prevented the rise of any significant sense of shared identity between East Bengal (Bangladesh) and the rest of Pakistan. Inter-regional mobility in India has been considerably restricted by the rise of linguistic patriotism. With the continued neglect of English coupled with the lack of any concerted effort to evolve a really effective link language, the sense of all-India nationalism has been adversely affected.

8. Finally, federalism is essentially a democratic union of consenting partners. To enable it to continue as a union of consenting partners, the constituent states must be assured full scope for articulation of their needs and

problems. As such, a functioning federalism cannot exist without a guaranteed multi-party democracy.

For successful operation of multiparty democracy it is necessary that the citizens possess a high degree of rule-consciousness. As Sawer put it, in any federation, the constitution is bound to be regarded, to some extent, as an obstacle to be got around rather than as the final word of the deity.

This urge to get it over with, or get around it, must be qualified with a certain anxiety, lest "getting around" disrupts the system. There must, in a federal society, be "a reasonable respect for orderly procedures in government, and a willingness to accept, with whatever grumbling, the compromises of plural politics and the advice of lawyers and decision of Courts on issues connected with the conduct of government" (Sawer, 1969, p. 135). But this rule-consciousness is a necessity for any type of orderly democratic government; not for federalism in particular.

5

HISTORICAL DEVELOPMENT AND SPATIAL FACTORS IN POLITICAL GEOGRAPHY

It is said that political geography is as old as the science of geography. Its study began along with the study of geography about two thousand five hundred years ago. Bat political geography, as understood by the ancient scholars, was very different from modern political geography. During the long period of its growth, the subject assumed several different forms.

During the early period (5th century B. C. to 17th century A. D.) the subject developed as a study of relationship of states with their natural environment. In the middle period (late 18th century to early 20th century) the main theme of political geography was the study of national power.' In the modem period the study of political areas (states) constitutes the core of the subject.

But it does not mean that as the new form was developed, the old one was abandoned. For instance, the environmental relations, which were the subject of speculation by the ancient scholars, were also studied by many political" geographers of the early 20th century. Similarly, the study of national power is still being made by some even today.

During the ancient period political geography was conceived as the study of the relationship of physical and political phenomena. The relation of states to the natural conditions of their territories formed the subject of speculation by the Greek geographers, historians and philosophers such as Herodotus, Plato, Aristotle and' Strabo. Herodotus, who was a historian, noticed an intimate bond between human groups and physical environment of their areas. Plato in his 'Republic' raised certain fundamental questions about the relationship of the state to its area. Aristotle wrote in his 'Polities' about the diversity of the people resulting from the climatic variations in different regions.

About the same time, politico-geographical concepts began to be formulated in India also. Kautilya's 'Arthashastra', written in the 4th century B. C. for the guidance of Emperor Chandragupta Maurya is an important treatise on the Indian state craft. Kautilya also set forth certain principles of internal and interstate relations.

Strabo, a Greek historian and geographer, in his study of the geographic laws of the Roman empire, concluded that the political unit for its proper functioning needed a powerful central government. He was impressed by the excellent location, climate and resources of Italy and recognised the importance of geographical factors in the development of states.

Among the Romans, Julius Caeser is known for his contribution to early politico-geographical thought. Because he gave due importance to physical environment, he succeeded in the expansion of the Roman empire.

During the Dark Age of geography, efforts of the scholars were directed largely to the religious topics and very little was, therefore, achieved in the field of politico-geographic thought. In the sixteenth century, however, the study of the relationship between natural and political phenomena was again taken up by Bodin, William Petty and Montesquieu. Bodin stated that national characteristics and political structure of states varied with changes in climate and topography. Montesquieu considered the effect of climate, topography and even continents and islands on people, laws and political systems.

William Petty tried to establish a science to deal with the relationships and interdependence of states and their territories. According to Goblet, Petty must, therefore, be regarded as the pioneer of political geography. More than half of his 'Political Arithmatick' is devoted to research on the interplay between states and their geographical environments and contains elements of true political geography. His theories dealt with the state's geographical sphere of influence, population density, capital cities and principle of intensity and distance as a factor controlling man's activities. He discovered two essential characteristics-position and extent of the geographical domain of the estate—and maintained that a dispersed territory is a source', of weakness. Thus Petty's ideas were a good basis of the emergence of modern political geography as an independent sub-field of geography. However, this early development of the subject remained obscure and an unnamed part of geographical thought (Goblet, 1956) for more than two centuries.

The Classical Period—Kant, Ritter and Ratzel

The celebrated German philosopher, *Immanuel Kant* (1724-1804) is regarded as the father of political geography, because he was perhaps the first to use the term political geography. Kant lectured on physical geography in the University of Konigsburg, Germany from 1756 to 1796. He held the view that physical geography was not only the basis of history but also of all other geographies, viz. mathematical geography, moral geography, political geography, commercial geography and theological geography. Thus it was Kant who centuries after the Greek, Roman and other scholars recognized political geography as a separate branch of geography. Kant's geographic ideas found very few followers outside Germany but within his own country there were several scholars who during the nineteenth century developed systematically many of the environmental concepts of earlier writers.

Karl Ritter

Quite a substantial part of the growth of political geography was due to the effort of Karl Ritter (1779-1859) and Freidrich Ratzel (1844-1904). Karl Ritter was Professor of Geography at the University of Berlin for several years. Often called the founder of modern geography in Germany, he first attempted to base conclusions of political geography on physical geography. He strongly believed in the deterministic influences exerted by nature. Ritter considered two sets of variables—physical environment (as it differs from place to place) and the developmental stage of each culture and propounded the theory of "organic culture". According to him human cultures are born, mature/and eventually die. He hold-that by understanding the laws of nature, a strong culture might be able to foresee and guide the destinies of the state.

Freidrich Ratzel

The most outstanding contribution to the development of political geography was made by Ratzel, who was for sometime Professor of Geography at the University of Leipzig in Germany. Like earlier scholars he also considered the question of environmental influences on the power of states. Ratzel's 'Politische Geographic' published in 1897, was the first systematic work on the subject, which rightly earned for him the title of the founder of modern political geography. He was essentially an anthropogeographer and made political geography a part of human geography. In his 'Anthropogeographic', published in 1882 and 1899 he made a thorough study of the earth and its people. He felt that an intimate relationship existed between the state and the soil. He tried to determine the significance of physical factors in relation to the state and ultimately admitted 'geographical determinum'. He noticed that the development of political geography was far behind that of all other branches of geography and that political science showed scarcely any trace of geographical influence. Ratzel, therefore, tried to accord a higher status to the subject by comparative investigations of the relations of states and the earth's surface.'

Theory of organic state : What interested Ratzel most was the growth of states." Basing his ideas on Ritter's theory of 'organic culture', he developed the theory of 'organic state', the concept of state as a spatial organism, composed of a portion of humanity and a portion of the earth. Ratzel was greatly influenced by the ideas of Darwin about natural selection, survival of the fittest and the need of species for space. Naturally he was led to compare the state with a living being. He held that like living beings, the state had the biological necessity of growth and for its expansion, it needed more space. Thus, according to Ratzel the state is involved in an endless struggle forgiving space or 'labensraum'. In his view, boundaries of a state were not fixed and with the advancement of culture, the state's geographical area was bound to grow.

Finally, Ratzel worked out seven taws of the growth of states:

(1) The space of states grows with culture;
(2) The growth of states enlarges following other manifestations of the spread of peoples and these manifestations must of necessity come before the state will grow;
(3) States grow through the amalgamation and absorption of smaller units;
(4) The frontier is the peripheral organ of the state and reflects, therefore, the strength growth and changes in the state;
(5) In their growth states seek to absorb politically valuable areas;
(6) The first impetus for territorial growth comes to the primitive state from without; and
(7) The general trend toward territorial annexation and amalgamation transmits the trend from state to state and intensifies the tendencies.

Though these ideas of Ratzel did not and favour with most foreign geographers, they were studied by his followers in Germany, who later developed the so-called German geopolitics.

The theory, of organic state, however, farms only a part of Ratzel's 'Politsche Geographic'. In the rest of the book relations with the growth and character of political units and their positions on the earth's surface, their size, boundaries, relation to the sea etc. are systematically and critically investigated (Taylor). In his writing, Ratzel stressed the importance of space (raum) and position (lage) as these determine the geographical value and ultimate destiny of every state. These are even today regarded as the two fundamental problems of political geography. The following quotation best explains the Ratzelian approach to the study of political geography. "In linking 'lage' (position) with 'raum' (space), Ratzel completes his study of the 'gebiet' (state's geographical domain) and rounds it off with an analysis of the limits of its extent, *i.e.* of its frontiers."

Several writers, in their evaluation of the work of Ratzel disapproved of the theory of organic state, because it ran counter to the idea of peaceful coexistence. For instance, the American writer and a pupil of Ratzel, Miss Ellen C. Semple (1911) eliminated the theory of organic state altogether in her restatement of the Ratzelian system. Some writers have even gone to the length of criticising him for his lack of objectivity and love for abstraction. It may, however, be admitted that Ratzel's 'Politsche Geographic' has provided a basic reading in political geography -in the twentieth century. It succeeded in paving the path of future political geographers for doing further work. At least his work delineated the essential parts of the subject in very clear terms and for many years guided the workers in the field of political geography.

Developments During the Early 20th Century—Political Geography into Geopolitics

In Germany political geography of Ratzel was appreciated more by statesmen, historians and political scientists than by geographers. It was because of the strong attachment of geographers to the orthodox branches of geography such as physical, economic and commercial geographies. For about two decades Ratzel's work stimulated little or no effort among German geographers to till the new field. Then the World War I started and its results brought into prominence several concrete problems which put forward a new demand for scientific political geography.

German geographers, Hettner, Supan and Schluter abandoned their previous positions and welcomed political geography into the innermost circle of scientific geography. The third edition of Ratzel's 'Politische Geographic' was perhaps in response to this new demand. Earlier in 1918 appeared the first edition of 'Political Geography' by *Alexander* Supan. Its preface assigned the subject a new importance to counterbalance the preponderate development of geography on; the physical side. Many books and articles on political geography by German geographers appeared in rapid succession. The book 'Guiding

Principles of General Political "Geography' by Supan was published in 1922. In the same year Walther Vogel and Arther Dix published their 'Politische Geographic' and three years later in 1925 appeared 'Politshe Geographic' by Otto Maull.

It is significant to note that all the works published by .the German geographers after World War I were directed towards restoring Germany to its prewar glory and power. It was political geography in name only, since the works hitherto published by German geographers were in reality German geopolitics, which emphasized the spatial requirement of the German state-Bowman (1927) remarked, "Political geography is still merely a term, not a science". He was perhaps right since the works of German scholars had nothing to do with the geographical science of political societies. Instead it was done with a view to justify the German territorial claims. As a matter of fact, their effort was largely directed towards the justification of Germany's demand for more space. As a consequence, for two decades (1925-45) political geography passed through a period of stagnation and decay in Germany.

Outside Germany, Ratzel's pioneering work influenced greatly the pro-German Swede, *Rudolf Kjellen* (1864-1922), who was Professor of History and Government at the University of Goteborg and of Political' Science at Upsalla, Sweden. Kjellen expanded the ideas of Ratzel and even modified slightly the concept of toe state as an organism by treating it as a conscious being on the same level as human being, capable of strong emotions. He believed that most important attribute of the state was power and that the power of maritime states would pass to the compact continental states, which would control the seas. He forecast that several great super-states would rise in Europe, Africa and Asia, and in Europe that state would be Germany. But Kjellen is perhaps better known for developing the new field of 'geopolitics', which term he coined and used first. His work had a great influence on the German geopolitical thought. He died in 1922 but his-ideas had taken root to give rise to the new discipline called German 'Geopolitics'. As referred to in the first chapter, it had nothing to do with geography in general and political geography in particular. As such, Kjellen rightly deserves a place among the contributors to the origin and growth of German geopolitics and not to the development of political geography.

The seed of geopolitics, sown by Kjellen, grew into a full tree largely through the effort of the German geographer, *Karl Haushofer* (1869-1946) He served in the German army and rose to the rank of a major general during the First World War. At the end of the war, Haushofer was appointed as Professor to teach political geography and military science at the University of Munich. Two years after the death of Kjellen in 1924 he founded the Institute of Geopolitics and started the publication of the 'Journal of Geopolitics'. Along with a band of geographers he developed several theories of the new discipline of geopolitics. He was influenced by the ideas of Ratzel and Kjellen as he accepted the Ratzelian concept of the organic state and the ideas of Kjellen on it. He applied the geopolitical principles for furthering the cause of the German nation. It is said that the work of Haushofer influenced the Nazi leader. Hitler, who was responsible for leading Germany to the World War II. Haushofer laid the foundation and built the edifice of geopolitics and therefore rightly deserves the title 'the father (founder) of German geopolitics'. Like Kjellen he contributed much to the development of geopolitics and very little to scientific political geography.

Development During the Last half century—American And British Contribution

So far we have discussed the contribution of German geographers to the development of political geography. Though modern political geography originated in Europe, very significant contribution to it has come from the American geographers during the last fifty years. It can rightly be said that the seed which germinated on the European soil, more correctly on the German soil, grew into a tall tree on the American soil. Isaiah Bowman was perhaps the first to write on political geography. His book published in 1921 was the first important work in the USA. Alaxander writes, "For the first time a geographer has compiled a volume in which the major problem areas of the time were assembled and analysed not only as individual studies but also in their relationship to the overall political pattern of the world". The most significant contribution to the nature and methodology of political geography .has, however, been made by Richard Hartshorne who together with Whittlessey and Valkenburg forms that trinity of American geographers who cleared the mist encircling the field of 'political geography' by their writings on the subject.

No American geographer has, perhaps made a greater contribution to the growth of political geography than *Richard Hartshorne*. As early as 1935 he attempted to establish a systematic framework of political geography which he based on the work previously done by German geographers. Writing in the pages of the American Political Science Review (1935) he discussed the progress political, geography had made upto that time and suggested the 'morphological approach' to the study of political areas. Fifteen years later in 1950 he came out in favour of the 'functional approach to the study of political geography. In 1954 he restated his view of political geography through a discussion of three other approaches to the study of political geography.

Thus Hartshorne made substantial contribution to the conceptual development of the subject. His three articles published in 1935, 1950 and 1954 may rightly be regarded as classical writings on the subject of political geography. Further work in political geography is rather impossible without going" through them. Hartshorne advanced an appropriate methodology, of political geography and delineated the scope of the subject. As a consequence, in all work which is now being done on political geography, his 'functional approach' is most commonly employed. If Ratzel is credited with giving .modern political geography its physical appearance towards the fag end of the 19th century, Hartshorne mast get the credit for giving the subject its present status and form. He has developed several theories and concepts and thus given an altogether new orientation to political geography. Besides, Hartshorne has also made-several studies on 'State idea', maturity of states and political boundaries.

During the period under review several books, articles and monographs on political geography have appeared in the U.S.A. Alexander has placed these writings under three categories: (i) Theoretical aspects, (ii) studies of political phenomena and (Hi) studies of specific political areas. In 1939 two excellent books, one by Whittlessey (The Earth and the State) and the other by Volkenburg (Elements of Political Geography) were published. While the former is a classical world regional text and contains several fundamental concepts, the latter deals mainly with the elements of political geography and the cycle theory of the evolution of the state.

In the early forties' a large amount of work of the American geographers was directed towards the repudiation of German geopolitics. Prominent among them are Isaiah Bowman (Geography vs. "Geopolitics) and George- Kiss (Political Geography into Geopolitics). These were aimed at distinguishing political geography from geopolitics. In the late forties and early fifties R. Hartshorne (Functional Approach and S.B. Jones (A Unified Field Theory of Political Geography) wrote extensively on the methodology and nature of political geography.

The studies of specific political features such as political boundaries, capital cities etc. are quite numerous. But it is not possible to list all the books and articles here. Mention may, however, be made of the books on boundaries by Boggs, Jones and Adami and articles by Richard Hartshorne, S.B Jones, N.J.G. Pounds, Eric Fischer, L.D. Kristof and J.O.M Broek.

A large number of publications in political geography made studies of specific political areas in the form of articles. But more striking feature has been the appearance of several world regional texts Noteworthy among them are : G.E. Pearey and R.H, Fifield (1948), H.W. Weigert and others (1957), L.M. Alexander (1964), L. Carison (1958), S.B. Cohen (1961).

Besides these books on regional political geography there are other books which deal with "the fundamentals of Political Geography written or edited by N.J.G. Pounds (1963), W.A.D. Jackson (ed. 1964), Harm de Blij (1967), R.E. Kasperson and J.V. Minghi (eds. 1969) and R.E. Norris and L.L. Haring (1980).

During the same period across the Atlantic in the U.K., the contribution of geographers to political geography was in no way less significant. Several outstanding professors have written excellent books and thus enriched the field of political geography. Among these are : A. E. Moodie (1949), W. G. East (1957),

O.H.K. Spate (1950), G.R. Crone (1957), J.R.V. Prescott (1972) and Richard Muir (1975).

SPATIAL FACTORS IN POLITICAL GEOGRAPHY

In the politico-geographical evaluation of any state, three geographical factors are of primary importance. These three (location, size and shape) are collectively known as spatial factors. 'Where is the political area', 'how large it is', and 'what shape does it possess' are some fundamental questions which must be answered. Not only they are to be described but their influence on the political life of the state must also be brought out fully, since the location, size and shape of the country influence the political organization and functioning of the state. Hence, each of the three factors must be carcfully considered in any politico-geographical appraisal of the state.

LOCATION

The importance of location and size was emphasized by Ratzel. He opined about three quarter of a century ago, that space and position determine the geographical value and ultimate destiny of every part of the earth's surface (political area or state). Their significance has not decreased even today.

There are three ways of expressing the geographical location of the political area :

(i) Astronomical or absolute location in terms of lines of latitude and longitude,

(ii) Vicinal or relative location with reference to neighbouring political units, and

(iii) Maritime or continental location, in terms of its relation to oceans and continents.

THE ASTRONOMICAL LOCATION

The astronomical or absolute location of a politico area is a geographical factor of great significance. As defined by lines of latitude and longitude, it is mathematically determined and cannot be altered. It may be subjected to change only when the political area expands or shrinks in size. But, as it rarely happens, me absolute location remains unaltered for along period of time. India's absolute location is expressed as between 8°N and 37°N and 68°E and 97°E.

The astronomical location of a state is significant, since it determines the climatic conditions which in turn influence the physical energy and economic and political strength of the people. Statements of Bodin and Montesquieu that people of the middle latitudes are best adapted to control politics and the weakness of the inhabitants of warm climates should make them slaves, have however lost most of their original validity. Modern writers like Huntington and Markhan have also expressed similar opinions. But it should not be taken to mean that political wisdom and power are the monopoly of the people of the temperate zone alone.

Nevertheless, it may be admitted that economic and political power of a state is related to the 'climate and through it to absolute location. Valkenburg writes, "climatic conditions are of marked importance to the activities of states and, therefore, to the formation of their political structure, and attitudes". He points out that the great powers of today are located within the north temperate zone. No doubt, economic productivity and health and vitality of people are influenced by the climatic conditions of the area. We may compare the excellent absolute location of the U. S. A. with the weakened absolute 'location of its northern neighbour, Canada.

The Vicinal or Relative Location

In politico-geographical terminology location of a state with respect to neighbouring states is known as the vicinal location or relative location. Location with reference to other states is probably the most important of the three locational aspects, since it influences the inter-state relationship as well as territorial security. Questions such as 'who are the neighbours', 'what is their number and character' are therefore significant in the political geography of every state.

However, the number is not as important a factor as the character of the bordering states. "The number of neighbours is a purely physical matter but then various characters involve human elements" (Valkenburg, 1963). Nevertheless, too many neighbours may render the vicinal location of a state complex and unsatisfactory. But no state can help it because it does not lie within its power to reduce the number of its neighbours.

Obviously, a state with several neighbours might be involved in a number of boundary problems, and during the war hostile and powerful neighbours may threaten its physical existence. For example, India's vicinal location may be said to be bad, because two of her seven neighbours are not very friendly. On the basis of sheer numbers the vicinal location of Russia with a large number of neighbour (12) is poor as compared to that of the U.S.A. which has only two neighbours. Though, no better example than that of the U.S.A. can be cited, yet the existence of one dozen neighbours presents to problem to Russia. Before the second world war, Germany with 11 neighbours had an extremely poor vicinal location from the point of view of national defence.

It may be seen that those states of the world which have only one or two neighbours are fortunate, because their vicinal location presents no problem. Certain island states have no neighbours since beyond their territorial waters lie the 'high seas'. The table given below presents a comparative assessment of the vicinal location of some countries :

Countries	*Number of Land*
China	13
Russia	12
Brazil	10
Germany	9
France, Zaire	8
Italy, Federal Public of Yugoslavia, Czech Republic, Niger, Mali, Tanzania, India	7
Laos, Turkey, Australia, Algeria, Libya, Chad	6
Bolivia, Argentina, Peru, Columbia, Hungary, Iran, Iraq, Saudi Arabia,, Burma (Myanmar), Ivory Coast.	5
Portugal, Denmark, Iceland, U.K.	1
Iceland, Philippines, Singapore, Malagasy, Sri Lanka, Newzealand, Australia	0

Coastal or Inland Location

The third way of stating the location of a state is to express it with respect to seas and oceans. Accordingly a state has either an inland (continental) location or coastal (maritime) location. This type of location has great significance, since it determines the character and interest of a state. The outlook of an inland state is markedly different from that of a coastal state and so are their activities and problems.

There are two types of inland states. Some are purely continental states with no sea boundary such as Afghanistan, Bolivia and Switzerland, to name only a few. Secondly, there are semi-continental states whose sea boundary is very small as compared to land boundary such as Zaire, or the nature of the long sea boundary is such that it generates no interest in the maritime activities. The coastlines of Russia have offered few maritime opportunities in its long history.

At present, 41 states of the world are purely continental (landlocked) and so are unfortunate. They have no free access to the seas and oceans and are forced to arrange for the movement of goods through the territory of a neighbour. As a result, their trade is small and if the neighbouring state is unfriendly, problems arise. Afghanistan is a case in point. For sometime Pakistan has been refusing Afghanistan the passage to the Arabian Sea. The economic development of such states is retarded, as they are remote, isolated and inaccessible.

Europe has the largest number of such purely continental states, followed by Africa and Asia. The continent-wise break-up of landlocked states is shown in the table below:

Europe	15 :	Switzerland, Austria, Czech Republic, Slovak Republic, Hungary, Luxemburg, Andorra, San Marino, Liechtenstein, Vatican, Bosnia-Herzegovina, Macedonia, Byelorussia, Armenia, Azerbaijan.
Asia	10 :	Afghanistan, Bhutan, Nepal, Laos, Mangolia, Kazakhstan, Khirgizia, Turkenistan, Tadzhikstan, Uzbekistan.
S. America	2 :	Bolivia, Paraguay.

Africa 14 : Niger, Chad, Central African Republic, Ruanda, Burundi, Mali, Burkina Fase, Zambia, Malawi, Uganda, Lesotho, Botswana, Swaziland, Zimbabwe.

On the other hand, inspite of access to the sea, some states are basically continental in their outlook. Russia and Canada are examples of such states. The long ice-bound coast lines offer fewer maritime opportunities.

Except the island states, such as the U. K. and Japan, all coastal states possess some degree of continentality in them, which varies from state to state depending on the type and character of sea boundaries, on the pull of the land as compared to that of the sea, and on the open and closed nature of seas that surround it. The observation of Valkenburg (1963) that "the more seas a nation fronts upon, the greater are its maritime opportunities and responsibilities" is therefore too general to command universal validity.

Taking the number of seas into consideration, coastal states of the world may be classified into five categories. A state may have one-sea location or two-sea or-three-sea multi-sea and island location. While Brazil, Belgium, the Netherlands, Rumania, Yugoslavia, Bulgaria, Burma and Thailand have one-sea location, the U.S.A., Mexico, Spain, Norway, India, South Africa have two-sea location.

Only a few states of the world have direct access to three seas. France, Canada and Turkey are examples of such states. The multi-sea location is vary rare and only large states possess sea frontage on several seas. Russia, by virtue of its being the largest state in the world, has a multi-sea location. But owing to the nature of its coasts, it has been essentially a continental state in its long history. The most extreme type of maritime location is the island or insular location. Examples of such states are Australia, U.K., Japan, Philippines, Indonesia, Sri Lanka, Cuba, Iceland and Newzealand.

SIZE OF THE STATE

Size, which refers to the areal extent of a political unit, is one of the three basic elements of any state, since it cannot possibly exist without a specific land area. Size may be stated in two ways—statistically and geographically. Statistically, it is expressed in terms of sq. miles or kilometers. But an entirely satisfactory method of expressing size geographically has not yet been evolved. Such expressions as 'a large state', 'a medium state' and 'a small state' often used to describe size in geographic terms are quite vague, since there are no uniform upper or lower limits of such size categories. Political geographers employ different terms for states of different sizes. A reference to size categories employed by Pounds, Valkenburg and de Blij.

Today, altogether there are 191 states in the world which range in size from the Vatican, the smallest with an area of 108.7 acres (less than 0.25 sq. miles) to Russia, the largest with about 65.7 lakh sq. miles. The largest state is thus about 26-5 million times as large as the smallest state. Very large states of the world in order of decreasing size are : Russia (66.7 lakh sq. miles), Canada (38.5), China (37.7), U.S.A. (36.1), Brazil (32,8), Australia (29.7), India (12.0), and Argentina (10.8). At the opposite end of the scale there are very small states, the micro states, all of which lie in Europe. These are : Andorra (1961 sq. miles), Liechtenstein (60), San Marino (24), Monaco (0.6) and Vatican (0.25).

Evolution of Size as a Spatial Factor

There is no denying the fact that size is a very significant factor in the power equation of any state. Without sufficient size no state can rank as a leading world power today. It is noteworthy that Russia and the U.S.A. are territorially very large and play very important roles in world affairs.

The size of a state is important for two reasons :

(i) A large size ensures abundance of resources,

(ii) it affords a greater amount of national security.

The importance of size was emphasized by Ratzel. "A large space maintains life", he stated categorically. The size still has the same importance as in the times of Ratzel. According to some it has greater importance today than at any time in the past.

One cannot definitely state that there is a direct relationship between area and resources of a state. The statement 'the larger the area, more are the resources', may not be absolutely true. Nevertheless, a territorially large state can usually be excepted to satisfy mow of its needs than a comparatively small state. A large state possesses a diversified natural resource base and thereby reduces its dependence on other states and enhances the chances of attaining self-sufficiency.

The large size is an important factor in the defence of a state. A large state, possessing "defence in depth", can successfully meet the challenge of the enemy state in any armed conflict. For instance, during the second World War, the erstwhile Soviet Union, due to its large size could easily "trade space for time" and could win the war waged by Germany. Even in the present age of nuclear warfare, sheer space is 'of great advantage. A very large state cannot be entirely devastated and several areas within the national domain may remain safe.

On the other hand, in a small country size is always a negative factor. Small countries have been unable to defend themselves in the event of an attack by large neighbours. Perhaps, small size was partly responsible for the defeat of the Netherlands, Belgium and even France in the second World War.

It may, however, be admitted that due to the development of nuclear weapons, even for large states, remoteness and depth have lost much of their meaning. Large size, in national defence, is no longer as important as it used to be in the past. Hence, there is a strong case of making a reappraisal of the size factor.

Large size is not always advantageous as in some cases it may not mean greatness. For example, Canada, Brazil, and Australia rank among the very large states of the world; Yet, none of these can lay 'its claim upon being a-major power. Unfavourable environmental conditions of large areas of these states tend to reduce the advantages of large size.

Great size can also pose serious problems in the fields of administration and transportation in peripheral areas. Without the necessary means to control it, large size is a big liability. In a very large state distances are great and governmental control tends to be less efficient in border areas where local interests, resulting from environmental differences, generate centrifugal forces and endanger national unity. Large size has, perhaps, been a major factor in the break-up of colonial empires in the 20th century.

On the other hand, some small countries attained considerable importance in the world, which was altogether out of proportion to their size. For example, the U.K. and the Netherlands rose to greatness because of their good location, favourable environment, and great industrial and commercial development. But small size could not sustain greatness for long. The decline of the U.K. bears ample testimony to this fact.

SHAPE OF THE STATE

Among the spatial factors, the snipe of the political area may appear to be the least important. The political value of Size has been questioned by only a few political geographers, but many have expressed doubt in the significance of the shape. Weigert has observed that "the shape is in many respects a haphazard characteristic without much significance". But in some cases it is important as it guides the internal development and functioning of the state.

A glance at the map of the politically divided world reveals that sovereign states of the world present a great variety of geometrical shapes. Although, no two states have the same shape, six categories of geometrical form may be distinguished:

(1) Compact;

(2) Elongated or attenuated;

(3) Fragmented;

(4) Prompted;

(5) Perforated, and

(6) Complex.

Compact Shape

Several states of the world are spread uniformly about their central areas and thus in shape are almost round or square. The ideal shape of a political unit is a perfect circle. But the shape of no county could be so regular as to, conform to a circle. A compact state possesses a shorter boundary than a non-compact state of comparable size. As a matter of fact, "the more compact it is, the shorter will be the length of its boundary in relation to its area".

The index of compactness may be expressed as a percentage or ratio. For instance, the length of the actual boundary of a state is expressed as the percentage of the minimum boundary, taken to be the circumference of the circle which would enclose the actual area of the State. Thus the index of compactness (%)

$$= \frac{\text{Actual boundary}}{\text{Minimum boundary}} \times 100$$

For example, in the case of a state, having an area of 8,000 sq. miles whose actual boundary is 400 miles and the minimum (circular) boundary forming the circle around the same area is 200 miles, the index (%) will be 200. It will be seen that the lack of compactness would be more as the percentage figure rises above 100.

Pounds has calculated the percentages certain statcs. They are : Uruguay (105), Rumania (137), Hungary (146), Switzerland (164), Belgium (167), Mexico (258) and Chile (310). Thus, Uruguay is nearly a compact state while Chile is not.

Valkenburg (1963), on the other hand, has made use of the ratio of the circumference of a circle containing the area of the country and actual boundary length, for expressing the deviation from the ideal circular shape. It can be calculated by dividing the length of the actual boundary (say 800 m) by that of the circular or minimum boundary (say, 400 m). Thus, a low figure will indicate that the shape is almost compact whereas a high figure will indicate the lack of compactness. For example, France's value of two indicates that its shape is nearly compact. Norway, with a value of eleven is an example of a non-compact state. Rumania, Hungary, Switzerland, Poland, Ethiopia and Tanzania are other examples of countries possessing nearly compact shapes.

The advantages of a compact shape are four-fold :

(a) A compact country has a shorter boundary to defend.

(b) In it cohesion of its various parts (national integration) is easily achieved,

(c) There is no difficulty in providing transport links between its various parts.

(d) The administration and control of peripheral areas pose no problems for the central authority.

Theoretically, compact shape may be advantageous in border defence. But since a strong country can defend lengthy frontiers, compactness is no great asset militarily. Finally, it may be said that "compact shape has probably a diminishing importance as the means of overcoming distance (and of defence) become available.

Elongated or Attenuated Shape

Elongation is another shape characteristic, too obvious to attract our attention. Some states are very long and at the some time also very narrow. Chile and Norway are two classical and striking examples of such states. In these cases, their surface areas are too much extended lengthwise in one direction. The ratio of length to width must, however, be fixed in order to decide whether a state has an elongated shape or not. De Blij (1967) putting it as 6 to 1, observes that "an elongated or attenuated state may, on the basis of Chilean example be defined as one that is, at least, six times as long as its average width". Thus, Norway, Sweden, Toga, Gambia, Italy, Panama and Malawi fall within this category.

Elongated shape is an odd shape and may sot be desirable. But whereas it has disadvantages, some advantage is also derived from it. Chile, which is 2600 miles long and only a little more than 100 miles wide, has no rail or road, running through the entire length of its territory from its northern border to its southern extremity. Norway has, however, an all weather road but no rail road. In both, heavy reliance has therefore been placed on a somewhat uncertain coastal shipping. In such states elongation has necessitated long transportation lines, involving huge costs and has created difficulties in the administration of remote and culturally diverse areas.

Physical diversity resulting from location in several environmental zones may, however, be advantageous. Chile, for example, possesses three distinct natural regions, as a result of elongation.

Fragmented Shape

Several states of the world have fragmented shapes since they consist of two or more parts, separated from each—other by either foreign land or high seas (international waters). Fragmentation physically more than obvious is, thus, brought by land and water. Most island states of the world are made up of big and small islands scattered over a vast area of the ocean. Indonesia, Philippines, Japan, Newzealand, Tonga, and Australia are good examples. A variant of the above is afforded by those states which consist partly of mainland (the main territorial unit) and partly of the offshore island or islands. Examples of such fragmented states are India, Malaysia, Italy, Greece, France, Denmark, Canada and Norway. Some states are divided into widely separated units, one of which is the main territorial mass, by intervening foreign land or water. Two modern examples of such states are the USA and Turkey. Other examples are afforded by Germany (1919-39) and Pakistan (1947-71). The above three fragmented shapes have been termed by Valkenburg (1963) as scattered, fragmented and broken.

As the non-contiguous or divided states are not of normal form, they are in a disadvantageous position. For smooth functioning as a political entity, the state should essentially have an uninterrupted contiguous territory. Due to the fragmentation of its territory, the state may face two-fold problem. In the first place, contact one part with the other may be difficult if not impossible. Secondly, integration 'of various parts may not be achieved easily because of cultural contrasts.

Pakistan of pre-1971 days is a glaring example. Separated by about 900 miles of Indian territory, one part was accessible from the other only by the sea route some 3000 miles long. Whenever Pakistan's relations with India became unfriendly, she had to face a number of defence and other problems. The unity .between the two wings could not be established and ultimately the eastern wing broke away to emerge as independent state of Bangladesh. Pakistan had to suffer due to its fragmented shape.

Thus in a fragmented state, whose territory is composed of main-land and islands, communication between the mainland and islands is not easy, if large bodies of water intervene; India, Malaysia and Indonesia exemplify such a case. But in inland countries which are fairly compact units such as the U.K., Japan and the Philippines, the lack of contiguity may not be felt seriously.

Prompted Shape

A state has a prompted shape if it shows one or two bulges, linear or otherwise extending from the main body of its; territory. Such bulges create serious problems of administration for the prompt state, because of the distant position from the central part. Bulges may be important areas either strategically or economically.

Zaire and Namibia in Africa, and Burma and Thailand in Asia are good examples of prompt states. Zaire has two proruptions—western in the form of a corridor to the ocean and the south-eastern in the minerally rich Katanga. In Namibia a long and narrow bulge (Caprivi's Fingure) extends right upto the Zambezi River between Angola and Zambia in the north and Botswana in the south, which sometimes back assumed great strategic importance due to the intention of South Africa to use it as a major military base.

In Asia, prorupt states of Burma and Thailand share a portion of the Malay Peninsula. Through it both countries extend southward for a long distance along the common boundary.

Perforated Shape

The fifth category of shapes is the perforated shape. The perforated state is one whose territory is pierced through by other state or the exclave of other state. The perforator state or the exclave is thus entirely enclosed by the perforated state. The best example is afforded by South Africa which is perforated by the state of Lesotho. Similarly, the micro state of San Marino perforates Italy. Since the access to the perforating state is possible only through the perforated state, the former is more inconvenienced than the latter and the perforated state has to suffer less.

It is not correct to say that the perforated shape is a bad shape, although, it may not be the normal shape as is the compact shape.

Compound or Complex Shape

Lastly, some states of the world have compound or complex shapes, since they possess more than one of the above mentioned shape characteristics. The continental USA, for example, is nearly compact, but at the same time it is also fragmented—Alaska is separated by the Canadian territory and the Hawaii by the oceanic waters. Italy combines three shape characteristics; it Is elongated, fragmented and perforated. Chile and Norway, although elongated, have fragmented" shapes due to some island possessions.

It may, however, be noted that only one of the several shapes is obvious and significant: For instance, in the case of Chile, Norway and Italy elongation is more significant than other shape characteristics. The compact shape of South Africa and France is more important than perforation in the case of the former and fragmentation in the case of the latter.

The complex shape possesses disadvantages of all shapes of which it is composed. It creates difficulties in state's unity efforts, administration and transportation. Only the latitudinal elongation, if present, results in climatic diversity which is economically beneficial.

The shape of India is complex and cannot be said to be good. The northern bulge in Kashmir, the southward peninsular extension and the north-eastern elongation add complexities to the country's shape. Besides fragmentation due to islands, the northeastern states of India are nearly cut off from the rest of the country. The northern boundary of Bangladesh comes so close to the Nepal border that they are accessible only through a narrow corridor (about 15 miles wide) of very difficult hilly terrain.

ECONOMIC ELEMENTS OF POLITICAL GEOGRAPHY

The states and the relations between them are greatly influenced by the economic elements of political geography. The economic elements are comprised of the basic natural resources of food, power, minerals etc., industry, trade and transportation. These six elements and their effect on the economic and political development of the state will be described and analysed in the following pages.

FOOD RESOURCES

As food supply is necessary for human welfare, it is the primary duty of each state to provide adequate food to its population. Human food consists of cereal and pulses, fruits and vegetables, milk and its products, meat, fish, eggs, sugar, oils etc. No state can be entirely self-sufficient in all food resources, since it is not possible for it to produce all types of food within its own territory. Nevertheless, the state has great responsibility for the supply of at least the staple food to the people much as cereals and pulses in the case of India. The self-sufficiency in food attained by a state is of course a great achievement but only few states such as the U.S.A. and France perhaps come closest to self-sufficiency in food. The states which are unable to produce the required amount of food must import it from the surplus countries. The industrial nations of Western Europe have long been importing food-stuffs from other countries in exchange for their manufactured goods. But poor and developing nations, whose purchasing power is low, are dependent on food supplies as loan or gift from other nations, particularly at the time of famine and drought. Such help is very often interpreted as interference with the political freedom of the nation involved. It has been said that hungry people may willingly forgo their freedom and accept foreign domination, whether communist or non-communist. "Food, probably the most important of the economic elements, has become a powerful political weapon in a world torn by conflicting philosophies and doctrines".

Too much dependence on food imports is undoubtedly a negative factor in the power of the state. The uninterrupted food imports are possible during peace time. But during the war food supplies may not be forthcoming since the transportation routes are blocked. Sufficient stock of food is therefore necessary in every state for use during the time of emergency. In should be the endeavour of each state to raise the production of food stuffs, so that dependence on imports is reduced and self-sufficiency so necessary for national strength is ultimately achieved.

India's Food Situation

At the time of India's independence, the food situation was extremely critical. In the beginning the country had very heavy dependence on food imports, as home production fell short of the requirement. But as difficulties in importing food specially cereals (wheat and rice) increased, it became necessary to increase foodgrains production in the country. The Government of India has tried to solve the food problem in a planned manner. In each of the Five Year Plans greater emphasis was laid on agricultural development as seventy percent of the people depend on this primary sector of the economy. The Green Revolution helped the Indian farmer in increasing the yield per acre of major food crops. More and more irrigation water was made available to the farmer, who was induced to use high yielding varieties of seeds, fertilizers and modern farming equipment. The result has been quite encouraging. The food situation is now much more improved than what it was at the time of independence. From a small amount of 55 million tons in 1949-50 the foodgrain production increased to 150 million tons in 1985-86. Whereas the population increased by less than two times from 36 crores in 1951 to 68 crores in 1981, the foodgrain production rose by about three times.

India is now almost self-sufficient in foodgrain production. The estimates put the production of foodgrains at 170 million tons in 1988-89. Although, India imported cereals and cereal preparations during 1983-86 she is now in a position to export some wheat and rice to other countries.

Power Resources

Power in one form or the other enters practically every single economic activity of the state and its availability and cost determine the economic future of the nation. Besides nuclear energy other sources of power are coal, oil, natural gas and hydroelectricity. As reserves of coal and oil are being depleted fast, governments have thought of using non-conventional forms of energy such as biogas, solar and wind energy. The extensive use of energy leads to economic prosperity which in turn enhances political power of the state. The home production of power and easy access to foreign soon are important factors in the evaluation of the power of the state.

No state is self-sufficient in power production. What to say of West Europen countries, Japan and India, even the U.S.A. is dependent on oil of the Middle East. Although, energy is in great demand in farm, factory and transport during peace time, its importance increases enormously during the war since easy availability of power, specially oil, is a deciding factor in the outcome of war. The importance of oil is great in modern economic development and a modern war cannot be fought without the use of large amount of petroleum and its products in various forms. "The failure of the German armies in their surge through Russia to occupy the Caucasus oil fields (only the Maikop field was occupied) and the later destruction of the German synthetic oil plant, were two great victories for the Allies. Their significance is evident by the fact that at the end of the war thousands of German planes ware captured on the air fields, unable to get away because they had no oil" (Van Valkenburg, 1963).

India's Power Resources

The coal position of India is satisfactory but the situation with respect to oil is rather alarming. The country has registered a five-time increase in coal production Since independence, from a mere 38 million tons in 1947, the coal production has increased to 154 million tons in 1985-86. According to the Geological Survey of India estimates of 1985, the Indian coal reserves were 1,55,900 million tons for coal seams of 0.5 metre and above in thickness down to depth of 1200 metres. The coal production is almost completely in the public sector, as coal mining was nationalized in 1972 and the Coal India Ltd. came into being in 1975. That the country now exports coal indicates that she is in a position to meet any sudden increase in the home demand of coal.

While self-sufficiency in coal has been achieved, the inadequate crude oil production has led to very heavy dependence on oil imports. Although, crude oil production increased from 2.5 lakh tonnes in 1950-51, to 290 lakh tons in 1934-85, this tremendous increase could not make India self-sufficient in oil. It is because along with increse in production the consumption has also been increasing (31 lakh tonnes in 1950-51 to 388 lakh tonnes in 1984-85). As the demand is more than the supply, India is compelled to import large quantities of crude oil from other countries, specially Iran. In spite of the efforts of Oil and Natural Gas Commission (ONGC), which is engaged in the work of exploration and exploitation of oil, the supply has fallen short of the growing demand. The production meets two-thirds of the domestic demand. The government has- tried hard to bridge the oil gap and succeeded in reducing it from 147 million tons in 1975 to only 98 million tons in 1984-85. Nevertheless, making the country self-sufficient in oil appears to be a distant possibility. Heavy dependence for this strategic fuel on foreign sources is considered to be a serious weakness of India.

The most convenient and versatile form of energy, the water power is next to coal as supplier of energy. Its demand has been growing at a rate faster than other forms of energy and as an inexhaustible source of power, electricity plays a crucial role in both agriculture, industry and transport. But hydroelectricity is in a backward stage of development in India. By 1985 only about 11.0% of the hydroelectric potential of 89.2 thousand MW had been developed. The installed capacity of hydroelectric power plants was 19.0 lakh kw in 1947, which rose to 114.0 lakh kw at the end of the Fourth Plan and to 155.0 lakh kw in 1985-86. The installed capacity has been increasing at a fast rate as a result of the construction of a number of major river valley projects, like the Damodar Valley cooperation (DVC), the Bhakra Nangal, Hirakud etc. Inspite of the phenomenal growth of Hydel power, its share in the total production of 170 million units of electricity in 1985-86 was only 30% (Coal was still the major source of electricity and contributed 67% of the total production. The nuclear plants contributed only 3%).

India is a peace loving country and is committed to the constructive uses of nuclear power, which has been developed on a modest scale. At the end of 1985-86 the, installed capacity of nuclear power plants was 12.7 lakh kw. The four neclear power stations are Tarapur near Bombay (Maharashtra), Rana Pratap Sagar (Rajasthan), Kalpakkam (Tamil Nadu) and Narora (U.P.). As mentioned above, nuclear power constitutes only 3% of the total generation of electricity in the country. It is heartening to note that the country has one of the world's richest deposits of thorium about 2.0 lakh tons, with 9% concentrate.

MINERALS AND OTHER RESOURCES

Much has been written on minerals and world politics in political geography and cognate disciplines. I here is no denying the fact that minerals have influenced international relations in all part of the world. History is replete with examples of interstate rivalry over access to mineralized areas, "Mineral resources may mean wealth specially to a small nation; however this situation may be temporary and may involve the danger of foreign control or interference". Since minerals provide one of the bases of industrial strength, they are much in demand in the world today.

No developed nation is entirely self-sufficient in minerals. 'A country may be sufficient in one or two minerals. But there may be some important minerals which need to be imported from other countries because of inadequate home production. Still there are minerals for whose supply the nation depends wholly on foreign sources. While procuring them may be easy during peace time, their supply may be completely stopped during the war. For example, the U.S.A. produces sufficient coal and iron ore. But for minerals such as mica, manganess, tin, and chromium she depends entirely on imports. As the lack of these essential minerals during the war inevitably proves to be a serious handicap, they rank high on the country's list of strategic minerals.

A striking feature of mineral resources is that they are distributed on the surface of the each much less regularly than cultivable soil. Nature seems to have been rather whimsical in scattering them through the earth's crust irregularly. This has given rise to international trade in minerals on a large scale. Another significant feature is that the mineral resources are exhaustible and incapable of being renewed quickly. Each deposit of a mineral if worked continuously will ultimately be over at any point of time. As a factor of state power minerals therefore, provide a great amount of uncertainty regarding their future availability.

India's Mineral Resource Base

Fortunately India is richly endowed with minerals, specially iron ore, manganese, mica, bauxite, limestone, magnesite, titanium and thorium. There are a few strategically important minerals which India does not possess at all or in sufficient quantity. They are' antimony, cobalt, mercury, molybdenum, nickel, tin, tungusten and vanadium.

Mining has recorded great progress in recent years. The value of mineral production in India (excluding atomic minerals) has increased from Rs. 1,112 crores in 1975 to Rs. 8487.3 crores in 1985. The share of fuel minerals (coal and oil) was however the largest being 87% of the total value, followed by non-metallic minerals 7% and metallic minerals 6%. The index of mineral production during 1985 (base 1970 = 100) was 240 as compared to 129 in 1975. The metallic minerals which India produces are iron ore, copper, chromite, manganese, lead, zinc and bauxite. India's non-metallic .minerals include limestone, dolomite, gypsum, apatite and phosphorite, mica and magnesite, Among the precious minerals gold and diamond are worth mentioning. India is now a significant exporter of iron ore, mica and manganese.

Resources other than minerals include agricultural fibres like cotton and jute, fibres such as wool and silk, and forest products nemely rubber, timber and wood pulp. These resources are important economic elements of political geography since they form a significant basis for the industrial development of a state.

INDUSTRY

"A highly developed manufacturing industry is the most conspicuous determinant of power. Every great power of modem times has been an industrial power". The distinction is often made between a developed country which is highly industrialized and a developing country which is not. As a matter of fact political power is synonymous with industrial power. All politically great nations have made phenomenal progress in manufacturing. Industrialization generates economic strength and is therefore considered to be a key to political power. The German Empire before the First World War was built on coal and iron. "The skilled exploitation of the great coal fields of Ruhr, upper Silesia and the Saar alone made possible the development of steel, chemical and electrical industries which established her as the first industrial nation of continental Europe".

Industrial capacity has been considered to be a good measure of national power. Several methods of measuring industrial capacity have been evolved. The amount of industrial production, specially of steel production, gives one such measure of the industrial strength of country. Another method of assessing industrial strength uses the percentage figure of industrial workers to the total population of the country. It may be inferred that higher the industrial production or the percentage of industrial workers to total population, greater is the industrial strength of the state. It seems that there is a direct relationship between industrial production or percentage of industrial workers and political power of the state.

Doubts have, however, been expressed as to the efficacy of these methods. The absolute or percentage figure of population employed in industries is not of much value since the effectiveness of the industrial labour varies from country to country because of unequal technological development. Moreover, uniform statistics are not available and some states even do not publish such statistics. Pounds has remarked that industrial production is no better index of political power for "it is difficult if not impossible to equate steel with textiles or chemicals with consumer goods... Furthermore, within each of the main branches of industry, how should one distinguish between producer of low grade steel such as prance and the producer of high-grade and alloy steel such as Sweden; between the chemical industry which produces fertilizers and that which turns out high explosives or plasticts or drugs. It is (therefore) doubtful that a rating of state on any absolute basis is possible".

India's Industrial Strength

India has emerged as one of the important industrial nations of the world on account of her most spectacular and impressive performance in manufacturing industries since independence. The industrial production has been showing an upward trend during the period of planned economic development. While the general index of industrial production was 109.2 in 1961 (base 1960 = 100), it rose to 184.3 in 1970. The trend in industrial production has been rising during the 1980's. The index of industrial production (base 1970-s 100) has increased from 154 in 1980-81 to 209 in 1985-86.

Not only has industrial production increased but its share in the net domestic production has also increased from 13.4% in 1970-71 to 15% in 1984-85.

Although, Indian economy is still largely agrarian, its industrial structure has grown much stronger since the time of independence. During the five year plans Indian industry has not only been expanded, but also greatly diversified. The expansion and diversification have been achieved by establishing new units of the existing industries and setting up several new industrial enterprises. For instance, India had only two major iron and steel units in the beginning of the plans. Now there are six major steel plants with an annual capacity of about 89 lakh tons. At the same time several new industries were set up. These are : electronics, engineering, tractor, locomotive, automobile, communication and power generation equipment, chemical fertilizer, plastics and drugs. The textile industry has been diversified by adding rayon and other units.

The expansion of the public sector is another land-mark of the industrial growth of India. Whereas in 1951 there were only five public sector undertakings with an investment of Rs. 29 crores, their number in March 1985 rose to 221 with an investment of Rs. 42811 crores. The public sector undertakings produce a large number of products such as steel, aluminium, copper, heavy and light engineering, fertilizer. Chemicals, drugs, locomotives, electronics, aircraft and ships. Having achieved self-sufficiency, it now exports its products of pig iron, steel, machine tools, rail coaches, industrial boilers, electronic instruments and power station equipments to other countries of the world. At present India's major industries are the metallurgical industries, mechanical and electrical engineering, chemical and allied, textiles and food industries.

TRADE

Economic self-sufficiency is an ideal condition which no state can ever achieve since it cannot produce everything in adequate quantity within its own territory. Goods not produced or produced inadequately have to be imported from other countries. No developed state is completely self-sufficient and free from dependence on imports. A state which produces a commodity at a high cost can import it more cheaply from another country, which has a large surplus for export. Thus, the necessity of international trade arises and all the states of the world are engaged in it.

The international trade gives rise to two problems. Firstly, the state with heavy dependence on import trade is always in a vulnerable position. For example. Great Britain has excessive dependence on food imports and therefore had to face great difficulty during two world wars because supply routes were blocked. Secondly, the state may face difficulty in making payment for the imported goods.

The payment is made by means of exports, by means of services performed by the importing country and by means of payment in gold, securities etc. However only a few countries such as the U.S.A. Russia and Japan can pay by means of services. The payment in gold etc. is made only in the time of emergency. It is therefore only by means of exports that payments are mostly made. It is for this reason that states try to promote their export trade.

There is no complete: freedom in international trade which is conducted by agreement reached between the governments. Each state controls the direction and volume of trade. It may refuse the export and import of certain goods. The state may regulate trade by quota system. It may impose tariffs usually on imports and sometimes on exports. But no state would like to lose revenue from duties imposed on trade. Some states, specially powerful ones, try to use their trade policies for fulfilling their political objectives.

The underdeveloped states are primary producers of raw materials from the forest, field and mine, while the developed countries import the raw materials and after converting them into manufactured products export them to industrially backward states. The developed states use their trade policy for political ends by pre-emptive buying and preferential tariffs. "Preemptive buying, in short, is the buying up of all or most of an available produce in order to prevent any one else from doing so.

This may be done in order to prevent a strategic material from falling into the hands of an enemy. This may also be done to place the seller in a dependent position politically". The state increases tariffs on the goods imported from another state for protecting its own industry. It may charge preferential tariff from a friendly state which it specially desires to favour. India has adopted the policy of conducting its trade with Nepal and Bangladesh on a "most favoured nation" basis.

After the Second World War the General Agreement of Tariff and Trade (GATT) was reached at Geneva in 1947. The agreement framed rules for the smooth conduct of international trade. It has succeeded in removing the artificial obstacles to international trade and the states now have wider freedom of conducting trade than before the GATT was signed.

India's Foreign Trade

Since 1950-51 the value of both imports and exports has been rising. The most striking feature of the foreign trade of India is that the balance of trade has always been, unfavourable signifying the excess of imports over the exports. India's exports during 1984-85 amounted to Rs. 16,855 crore and imports to Rs. 17,173 crores, leaving a trade deficit of Rs. 5,318 crore. In 1988-89 India established an all-time record of Rs. 20,281 crore in its export trade. But as the imports amounted to Rs. 27,693 crore, the trade deficit further rose to Rs. 7,412 crore. This is not a very happy situation. The ideal situation is one in which the imports and exports are almost equal. But it is very difficult to achieve that ideal. The national trade policy, announced in April 1985, therefore, aimed at boosting exports and encouraging effective substitution of imports to reduce the trade deficit.

Pattern of Export-Import Trade

Whereas at the time of independence, India exported raw materials and imported manufactured goods, she is now an important exporter of manufactured goods. The exports have not only increased in value but have also been diversified.

The important items of exports in order of decreasing value (provisional) in 1985-86 were gems and jewellery, readymade garments, tea, machinery and transport equipment, iron ore, leather and, leather manufactures, fish, cotton, chemicals, jute and spices. The chief items of import trade in 1985-86 were petroleum and petroleum products, machinery and transport equipment, iron and steel, pearls, precious and semi-precious stones, fertilizers, chemicals, vegetable oils, non-ferrous metals, artificial resins, plastic materials, paper and paper board.

For the speedy growth of Indian economy imports of petroleum and its products, specific type of machinery spares and components, iron and steel, fertilizer, non-ferrous metals and other industrial raw materials are required and India will continue their import for the time being. Significantly the first seven items of imports make up over 70% of the total imports.

Direction of Foreign Trade

Although, India has trade ties with almost all developing and developed countries of the world, three major trading partners are the U.S.A. Russia and Japan. The dominance of the U.S.A. in both import and export trade is most striking indeed.

Besides the above three, other important countries with which India has trade relations are Germany, U.K., Belgium, France, Italy, the Netherlands and other E.C.M. countries, Saudi Arabia, U.A.E., Iraq and Kuwait in West Asia and Iran, Singapore, Hongkong and Australia of ESCAP (Economic and Social Commission for Asia and Pacific). One striking feature of the foreign trade of India is the unfavourable balance of trade with almost all legions of the world.

Hence, it is necessary to increase the value of exports so that the trade deficit which at present characterizes India's foreign trade is wiped out expeditiously.

TRANSPORTATION

"Transportation is a major factor in the evaluation of nation in peace as well as in war". There are two aspects of transportation. One is external and is concerned with the international trade which is by land (rail and road), water (ocean, river, lakes and canals) and air to a very limited extent.

The other is internal which plays a very important role in the economic and political development of the state. Transportation is significant in national defence, political unity and economic development of the state. Thus an efficient means of transportation of men and material from one place to another is a necessary element of the national power.

In war, defence of a country is possible with the facility of transporting army personnel and military equipment from one part of the country to another. The strategic roads and railways are therefore, necessary to strengthen military power of a nation.

The unity and strength of a nation depend to a large extent on the lines of transport and communication. Many states of the world possess a large network of roads and railways radiating from the national capital to all parts as in India, France and Russia. A well developed transport system is considered necessary for the political unity of the state.

New Delhi is very well 'connected with all parts of India by roads, railways and airways. As a matter of fact the unity of British India was a creation of a large railway network. "Before the communist Revolution, a rail road was built from Moscow to the Ural Mountains and across Siberia to Vladivostok. Since then this line has been double tracked, supplemented by a net of branch lines and made more effective". The trans-continental rail lines of Canada and the USA have played a major role in uniting the eastern parts with the western sections of each of the two countries.

Without an efficient network of transportation lines raw materials from the field and mines cannot be assembled at the factory sites and similarly manufactured-goods cannot be brought to the market places. The economic development of a country cannot be achieved with a faster rate without an efficient network of transport. During drought and famine food supplies can be transported to the affected areas in no time.

The railways are supplemented by roads and inland waterways in an overall transport system of a country. Roads can be built in the mountainous areas where rail lines are difficult to construct. The water transport is relatively cheap and suited to the movement of bulky goods.

It is possible in all countries where there are navigable rivers, lakes and canals. India has an ambitious plan to develop the Ganga river as an inland waterway from Haldia near Calcutta to Allahabad. In European countries such as France, Germany and Russia a large number of canals have been cut to interlink the navigable rivers to make inland water transport more effective. The Great Lakes of North America are the finest example of transportation by lakes.

Transportation in India

There is no exaggeration to say that railways are the lifelines of India. They are no doubt the chief means of transportation with a total route kilometerage of 61,850 (March, 1985), out of which 6,325 km are electrified. The Indian railways are the largest in Asia and fourth largest in the world after the U.S.A., Russia and Canada. For administrative purposes railways have been divided into nine zones :

Central, Eastern, Northern, North-Eastern, North East Frontier, Southern, South Central, South Eastern and Western, It is needless to say that railways have played a major role in national unity, defence and development of India.

The road network of India is one of the largest in the world. The total length of roads, both surfaced and unsurfaced, was 15,54,204 km, in 1982-83. Indian roads are classified into national highways, state highways and rural roads including village and district roads. "Roads are being developed in rural areas under the minimum needs programme, the objective being to link all villages with a population of 1500 and above and 50 percent of the villages with a population of 1000-1500 with all weather roads by 1990" (India 1986).

The roads have strengthened military defences considerably in the remote and inaccessible areas. "The Border Roads

Development Board was set .up in 1960 for accelerating economic development and strengthening defence pre-paredness through rapid and coordinated improvement of road communication in the north and north-eastern border areas".

In India river transport was much developed before the coming of the railway. As a matter of fact, the decline in river traffic was caused by the use of railways. Further, the irregular water depth and shifting channels due to the seasonal character of rainfall and diversion of river water through canals for irrigation led to the decline of river navigation.

Although, the country has 5,200 km. of navigable stretch of major rivers, only 1700 km, are being used. The important navigable rivers are the Ganga, the Brahmaputra, the Godavari, the Krishna, the Mahanadi, the Narmada and the Tapti. The Allahabad-Haldia stretch of the Ganga-Bhagirathi-Hooghly river system will be developed under the National Waterways Act of 1982.

The navigation has been started in the Haldia-Farakka section. But river traffic upstream of Farakka will be possible on the commissioning of navigation locks at Farakka. When completed, the Haldia-Allahabad inland waterway will be the longest in India.

The canal navigation is not very popular in India. The length of canals in which navigation is possible is 4300 km. But only 485 km. is suitable for mechanized craft (331 km. are actually in use). The Buckingham canal in the coastal area of Andhra Pradesh is the most famous of all the canals. The Badagana-Mahe and Cochin-Quilon canals connect the lagoons on the sea coast 'of Kerala. The air transport in India is operated by two public sector corporations named Indian Airlines and Air India, which were established in 1953.

While the former operates air services within the country between New Delhi and all the State capitals and other important towns and cities, the latter operates services to some fifty-nine countries of the world. Vayudoot, the third airline was incorporated as a Private Ltd. Company in January 1981 "to connect inaccessible areas of N.E. region and stations which are not covered by Indian Airlines and are important centres of trade, commerce or are important from the point of view of tourism". The last but not the least are the pipelines which transport both crude and refined petroleum and natural gas over long distances.

In India pipelines have been laid in Assam, Bihar and U.P. to facilitate domestic transport of oil. Of late natural gas is being transported from the gas fields to the fertilizer factories.

The Hajira-Bijaipur-Jagdishpur (HBJ) gas pipe line (1730 km.) built by the Gas Authority of India Limited (GAIL) carries gas from Hajira in Gujarat to the gas based fertilizer plants of M.P., Rajasthan and U.P.

6

GOVERNMENT AS AN AGENT OF GEOGRAPHICAL CHANGE

INTRODUCTION

Political geography has traditionally been involved in the study of two sets of problems. One was concerned with the study of the influences of physical environment on political activities, particularly those relating to man's political organization of space. The other, related to the analysis of how political organization of space modifies the landscape of the area upon which the given political organization impinges. However, in practice, most studies concentrated on the former, and for long, the second perspective remained almost completely neglected. A pointed attention to this problem was drawn by Derwent Whittlesey (1890-1956) (1935) in a paper published in the *Annals of the Association of American Geographers*. Whittlesey stressed that "Political activities leave their impression upon the landscape, just as economic pursuits do", and that "deep and widely ramified impress upon the landscape is stamped by the functioning of effective central authority", which today is "the function of the nation state".

In this paper, Whittlesey demonstrated various ways in which the functioning of effective central authority brings about modifications in the cultural landscape. He divided his treatment of this theme into four major sections:

(i) Expressions of security

(ii) Special features of boundaries

(iii) Expressions of government activity, and (iv) The effects of the legal system. Here is a summary of Whittlesey's paper.

THE IMPRESS OF EFFECTIVE CENTRAL AUTHORITY UPON THE LANDSCAPE

Expressions of Security

Security of life and property is the most far-reaching product of effective central authority. In a state with assured internal and external security, settlements grow freely across the national space, wherever suitable facilities exist. In the countries of the New World which were settled in the age of effective central authority, farmsteads are dispersed, and even where the agricultural mode favours clustering—as in irrigated districts and on plantations—the farmsteads line up as string towns along the roads so that each may be centrally located in the midst of its farmland. This is in sharp contrast to the case in the long-settled countries of the old world where, the houses may cluster, in tight knots, and many may be perched on defensible hills or protected by water, and inconveniently removed from the farm land.

> In countries which were settled in eras of insecurity, defense points have now been—abandoned in favour of more convenient sites. Crowded hills are deserted for open plains—Where artificial defenses were formerly maintained, moated granges and villages free themselves for expansion by filling their encircling waters by their encircling walls. Where the agricultural system cogently favours scattered habitations, isolated farmsteads appear, concomitant with security.

Urban settlement also spread out following guaranteed security. The spacious residential suburb is a product of the age of security as much as a consequence of fast transport.

Special Features of Boundaries

When central authority is ineffective, border districts, though legally subordinate, in practice remain at liberty to carry on their life unhampered by governmental restrictions. The frontiersmen can work out a mutually beneficial economic relationship with the border communities across the international boundary. The inhabitants of the border regions feel closer to the border communities of the neighbouring State than to people elsewhere in the country to which they legally belong. With the rise of effective central authority, the border regions are subordinated, their links with regions outside the national space are severed, and through the development of effective network of transport and communication, the border districts are fully integrated into the national mainstream.

Boundary displacements create problems and leave political and economic scars.

> Routes and other communication patterns are frequently altered after a boundary displacement; at first certain connections are closed or so restricted by inspection at the border that they fall into disuse; then new connecting lines, suited to the new alignments of territory, are built.

Expressions of Governmental Activity

In Specified matters, the central authority undertakes administration for the entire area of the state. This leads to uniformity in cultural landscape even where the natural landscape may be diverse. The most conspicuous result of the central administrative functions of the State are the public buildings of uniform function and form. In the new countries in North America, routes and land lots were sketched upon the landscape antecedent to settlement. Governments often stimulate migration into newly acquired areas by offering larger land holdings and subsidised infrastructure. Modifications occur as a result of government policy to promote development in backward regions by transferring resources from the richer areas by means of taxation.

Laws Resulting from Landscape Modification

Trade restrictions leave their mark on the economic landscape. The countries which fence out foreign manufactured goods, give rise to new industrial districts, while the ones allowing unrestricted entry to foreign manufactured goods would stamp out indigenous industries. Tariffs and subsidies in agricultural production lead to economic patterns different from what would have existed otherwise. Large acreage of wheat in France, intensive cultivation of tobacco and beet sugar in nearly every country in Western Europe, sugarcane in Hawaii, Puerto Rico, Philippines and Louisiana (USA) are some examples.

In the Old world, every major social revolution produced its own crop of new laws which affected land holdings. In general, such laws were calculated to prevent the rise of large estates (and break the existing ones), to restrict holdings to small acreage, and to limit agricultural development to those modes in which small holdings prove profitable.

Thus in many parts of Europe, subsistence farming replaced commercial farming, following the First World War. Small holdings in Europe favoured live-stock raising instead of grain production, since the small proprietor can pay careful attention to his animals, but cannot afford costly machinery required for grain cultivation on a commercial scale.

Since the publication of Whittlesey's (1935) paper, the theme of the impress of effective central authority upon the landscape has been considered to be worthy of fruitful research in geography. But because of the retarded nature of political geography, the subject received very little attention. The few published studies were by and large, limited to the examination of the landscapes of interstate boundaries in federal States.

By mid-sixties interest in the subject had greatly increased, so that Cohen (1966), thought it appropriate to forecast the emergence of a "geography of policy". Prescott's book *Geography of State Policies* (1968) appeared soon after. But generally, geographers were engaged in writing sermons rather than making a genuine effort to resolve problems.

As a reviewer noted, the two key chapters in Prescott's book consisted "largely of suggestions for future research, rather than [even] a synthesis of research completed" (Roepke, 1970). For the first time in 1971, an attempt was made by Dikshit to evolve a framework for the study of governments as agents of geographical change (Dikshit, 1971b). The basic ideas of the proposed framework are summarized below.

TOWARDS A FRAMEWORK FOR THE STUDY OF THE EFFECTIVENESS OF CENTRAL AUTHORITY UPON THE LANDSCAPE

Form or Structure of Government

In the discussion of government influence on the location of human activities attention is seldom drawn to the fact that it is not the government *per se* but rather the *form* or *structure* of government that is of fundamental importance in determining its effectiveness as a locational agent. Evidence suggests that in the study of the impress of the central authority, it is far more significant to know whether a government is unitary or federal, democratic or non-democratic, rather than whether a particular country has an under-developed (*i.e.*, poor) or developed (*i.e.*, affluent) economy.

Though the United Kingdom and Australia are among the so-called developed economies and both are inhabited by people largely drawn from the same stock, the nature of central authority in the two is very different. Consequently, the effectiveness of governmental influence as an agency of geographical change is also different in each case.

As discussed in Chapter 8, broadly speaking, there are two different systems of governmental organization: unitary and federal. In the unitary system, there are only two levels of political authority: local (the grass-roots level) and national. Local authorities are assigned a limited role of managing local administration under the broad outlines set by the national government, which remains the sole legislative authority, empowered to formulate policies and oversee economic and other developments for the entire country. As the supreme authority, in matters other than purely local ones, remains the same everywhere, the location of the different kinds human activities in such a State is likely to be regionally balanced, in so far as the spatial distribution of resources (physical and human) permit.

In sharp contrast, in the case of federal political systems, between the national government at the centre and local governments at the grass-roots level, there are state governments with constitutionally guaranteed autonomy in a number of important fields, including economic and cultural. To enable the central government to plan economic activities in fields other than those in which the constitution has empowered it, it is necessary that a majority of the member states should agree to delegate the requisite authority to the centre. This way, the centre can formulate policies in areas other than those covered by its original competence. If a consensus is not reached or the state governments in the federation are controlled by opposition political parties with conflicting economic philosophies, there may in a federal State, occur sharply contrasting patterns of economic development and social well-being.

It is this quality of legislative authority in federalism that tends to create what a Canadian statesman termed as "cadillac areas and cart areas" in the same country (cited in Mallory, 1965, p. 13). Thus, in a federation, distribution of economic resources, man-power, and the availability of markets, do not often decide the spatial pattern of development. Rather, it is regional politics born out of the national system of governmental organization that in the ultimate analysis determines the pattern and direction of development.

Governmental Process

No less important than the administrative structure is the nature of governmental process (that is, whether it is democratic or totalitarian). The United Kingdom, Sweden and China are all "functionally" unitary in their governmental structure. But because of the multi-party democratic process of government in the first two and a one-party "democracy" in the third, the influence of government on the location of economic and other types of activities is very different. British political leaders, with an eye on the next general elections cannot afford to adopt any planning measures which completely disregard the prevailing sentiments of the people, even if the leaders themselves consider these measures to be in the national interest. But a country like China, because of its one-party democracy, can very well afford to do so by virtue of its capacity to turn the entire economic and military effort into a certain desired direction, without fear of argument, strikes, or lack of public support.

Stage of Resource Development

A third decisive element in determining the effectiveness of central authority is the stage of the country's resource development. On such a basis States may be classified into three broad categories: namely,

(i) States with developed resources-use at the present stage of technological development;

(ii) States with developing resource-use (usually densely-populated and less-industrialized countries such as India); and

(iii) States where the resource-use is still at the pioneer-fringe level as, for example, in Australia. The more developed the resource-use in a country, the greater is the need for co-ordination of economic activities in the nation. Hence, the State's economic and political organization is likely to be more organic. Under these conditions, the central authority will be more effective in bringing about uniformity in national standards and a balanced regional development across the country. Similarly, the less developed the resource-use, the less is the necessity for centralized planning and, therefore, less effective would be the central authority in determining the location of economic and other related activities.

Among federal States, countries at the developed and developing stages of resource-use are likely to be more integrated and organic in their governmental organization. Whereas, in States still at the pioneer-fringe level of economic development (focused on primary production), conditions similar to those of the nineteenth century United States may still prevail, and laissez-faire philosophies of economic development may still be in vogue.

As Linge (1967) demonstrated, in Australia for a long time, each state had functioned virtually as an independent economic unit and economic competition between the states was very great. Although "government decisions at the federal level, such as tariff protection, have had a considerable impact on the Australian industrial structure in general, the internal distribution results almost entirely from the policies of the state governments" (Oxtoby, 1970, p. 464). It is, no doubt, true that the "Commonwealth of today is clearly a more activist and initiatary organism than the grand co-ordinator—envisaged Ay most of the Founding Fathers" (Spate, 1968, p. 187). But all that the increased centralization of power in Australia has succeeded in doing is to make "the states as developmental agencies and the Commonwealth as a disbursing authority" (Rabinson, 1961, p. 13). The point is significant: in Australia the Commonwealth is still primarily the "disbursing" and not the "developmental" or planning authority. In federations with more developed resources-use, such as the United States or India, the planning functions of the central government are far more important though they can nowhere be as effective as in the case of the unitary States.

Cultural Geographic Base as a Modifying Element

The actual functioning of governmental authority is greatly modified by the regional distribution of ethnic elements within the State. In multi-ethnic federations, for example, spatial arrangement of ethnic elements in the population may greatly modify the effectiveness of the central authority and its influence, in bringing about a spatially uniform pattern of development. The Canadian centre was originally far more powerful than the American, and the Indian centre more powerful than the Australian, but the facts have changed since. In both of these multi-ethnic federations, regional concentrations of clearly identifiable cultural elements have tended to create miniature linguistic and/or cultural homelands, which are averse to any central intervention in state (*i.e.*, provincial) affairs.

This has often led to the control of regional governments by region-based political parties, pursuing divergent political, economic, and social philosophies. Constitutionally guaranteed regional autonomy in a number of important socio-economic fields under the federal set-up, may give rise to the evolution of very disparate patterns of development. In most unitary States, however, such complications would not arise, for here "the sociological unit of the region is [often] not powerful enough to demand social account" (Boudeville, 1966, p. 55). Wherever provincial sentiments are relatively strong, the matter is resolved with some special arrangements, but seldom with the grant of regional autonomy. Thus, the central government, under a unitary set-up, remains the sole planning and developmental authority and its influence on the location of economic and other related activities remains almost uniform (within the constraints imposed by physical and other factors) throughout the entire territory of the State.

The Hypothesis

We may therefore conclude that; the effectiveness of the central authority as a landscape modifying agency in any given State depends on the governmental structure, process, and the stage of resource development. But the underlying cultural-geographic base may greatly modify the degree of governmental effectiveness as an agency of geographical change.

Discussion

To remove any doubts that might be entertained against the theory proposed in the foregoing discussion, it may be helpful to present a brief comparison between the nature of regional planning in the United Kingdom vis-a-vis Australia. Asking himself what kind of machinery does the British regional planning system have, Self (1967, p. 9) wrote:

1. The whole machinery is advisory only. The civil servants on the boards are primarily concerned with their departmental duties, and their loyalties are vertical (to departments) rather than horizontal (to-the regional boards).
2. As departments have different views, concrete questions will tend to gravitate to the Whitehall, to be settled by the 'usual machinery' (including the Cabinet or a Cabinet Committee if the issue is very controversial).
3. The regional advisory councils have no elective roots or staffs of their own, and occupy a generally inferior position to the boards of civil servants. They will have to struggle to exercise any political leadership at the regional level.

To sum up, the "whole machinery (of regional planning in Britain) is advisory only", and the civil servants, who are the moving force behind this planning, have loyalties that are "vertical (to the departments) rather than horizontal (to the regional boards)". In view of this "concrete questions...tend to gravitate to Whitehall" which remains the sole planning authority for the entire country.

Thus, what pattern of development the nation and its regions will adopt is decided by the central government. Detailed implementation and administration of these decisions are left to the centre-appointed and region-based officers who, in consultation with the concerned regional planning board proceed in a manner better suited to the local needs. The planning boards are, therefore, at best region-based but centrally-controlled agencies, without any autonomous decision making powers or financial resources. This rules out the possibility of any intense regional competition resulting in zero-sum games of the type often played by the state governments in Australia. Regional planning throughout Britain, remains under the control of centre-appointed officers whose loyalties are vertical to the national government.

Thus, the scope for contrasting patterns of regional development is greatly minimized. In sharp contrast to this, the loyalties of the regional planning officers in the Australian states are "horizontal" to their respective state governments. Here, the co-ordinating power of the Commonwealth remains limited because of the country's federal constitutional structure, its democratic governmental process, and its pioneer fringe economy. These together work to limit the scope for central co-ordination and control. Chart 6.1 (based on Dikshit, 1973) illustrates the framework discussed above.

India : A Case Study

India has a peculiar colonial heritage as a single colony with a unified administration for almost a century. Unlike other post-colonial federations of the former British Empire, India developed into a federal State with strong unitary features. After independence, she emerged "as a federation in which the paramountcy which the British Government had over the Indian States has been taken over by the Union Government and applied to all its units" (Santhaham, 1960, p. 13). For instance, Article 300 of the 1950 Constitution reads:

The Government of India may sue and be sued by the name of the Union of India and the Government of a State may sue or be sued by the name of the State and may, subject to any provisions which may be made by Act of Parliament or of the Legislature of such State enacted by virtue of powers conferred by thc Constitution, suc or bc sucd in rclation to their respected affairs in 'the cases of Dominion of India and the corresponding Provinces or the corresponding Indian States might have sued or been sued if this Constitution had not been enacted.

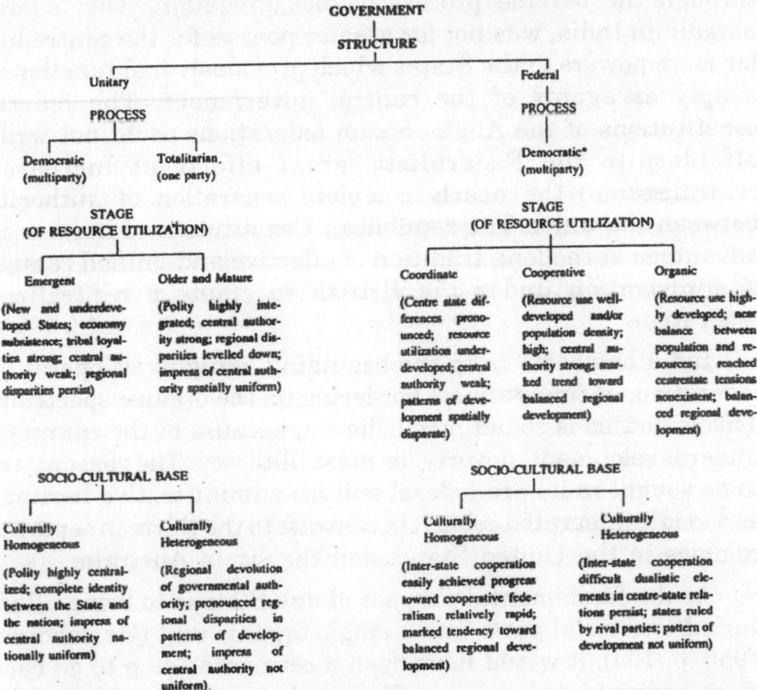

Fig. 6.1 : A Model for the study of Government as an agent of geographical change: The impress of effective central authority-upon the landscape, as a function of structure and process of governmental organization and the stage of resource development (as Modified by the Sociocultural Base).

Behind the republican Constitution of India stands the legacy of a long and unified British rule out of which the Indian federation evolved. Provisions of similar nature may be found in the *Government of India* Act 1935, Article 176 (i); the

Government of India Act 1915, section 32; as well as the *Government of India Act 1858* which reads:

> All persons and bodies politic shall and may have and take the same suits, remedies and proceedings, legal and equitable against the Secretary of State for India as they could have done against the said (East India) Company.

In contrast to the three Anglo-Saxon federations, the Indian federation was born not through a process of aggregation but through the reverse process of disaggregation. The federal bargain in India, was not for greater powers for the centre but for more powers to the States which previously had functioned simply as agents of the central government. The federal constitutions of the Anglo- Saxon federations could not write off (despite the Federalists' great efforts at increased centralisation) the nearly complete separation of authority between the units. The republican Constitution of India took advantage of the long tradition of effective and unified central administration under the British to create a centralized federation.

India has been, from the beginning, a highly co-operative federation, at times almost bordering on the organic spectrum. This is not, as is sometimes believed, because of the country's underdevelopment, poverty, or mass illiteracy. The reasons are to be sought in its pre-federal politico-administrative heritage as a single integrated colony, in contrast to the thirteen separate colonies in the United States and the six in Australia.

As the fundamental purpose of any State is to bring all its varied territorial parts into a single organic unit (Hartshorne, 1950, p. 104), it would have been a retrograde step to go back to co-ordinate federalism. Thus, what others achieved by evolution, India received by way of heritage.

All through the British period, the whole of the subcontinent had formed a single economic unit, and the flow of goods and capital had been free throughout the country. Economic investments in the country were located, in the areas of maximum profits, irrespective of the provincial boundaries. As the constitutional provisions after independence guaranteed a common market, the earlier pattern of economic development continued, irrespective of the new set of internal political boundaries.

Announcing its first Industrial Policy Resolution in 1948, the republican Government of India recognised four basic groups of industries: (i) *the State monopolies,* consisting of arms and ammunition, atomic energy, and railway transport (to be exclusive spheres of the central government) (ii) *the basic industries,* which consisted of coal, iron and steel, aircraft, ship-building, petroleum, and the manufacture of telephone, telegraph, and wireless apparatus (additional manufacturing could be developed only in the public sector though the government could secure the help of the private sector if necessary), (iii) *the public sector industries,* consisting of industries under government control and regulation, such as automobiles, tractors, machine tools, and cement, (iv) *the private sector industries* not subject to government regulations and controls (although the government retained the right to enter any industrial field not covered under groups...(ii) and (iii)) Thus, under the terms of its stated competence, it was envisaged that the central government would play the decisive role in the location of industrial establishments across the country.

The Industries (Development and Regulation) Act, 1951, gave further powers to the centre to control industries in the private sector. Each industrial establishment using power, and employing over fifty persons (or not using power, but employing over 100 persons) had to be registered with the government. The government could set conditions concerning the location of industries through its registration and licensing policy, and it could reject applications on the basis of unsuitable locations (Government of India, 1953, p. 4 and p. 103).

A new Industrial Policy Resolution issued in 1956, besides other things, called for the reduction of differences in the levels of development between various regions, so that benefits of industrialization could reach the whole country. With this objective, the establishment of Industrial Estates was proposed, to provide facilities for small and medium-size industries in certain under-developed areas. Since 1951, the Government of India has launched successive Five Year Plans envisaging definite policies of industrial location. The Third Five Year Plan (1961 to 1966) called for the establishment of new industries away from large and congested urban centres, with a view to initiate new regional growth points. (Government of India, 1961, pp. 454-456). Thus, as Irvin Roth (1970, p. 386) correctly put it,

> the government is aware of the need for an industrial location policy and the main stated objective is to place industry within the country in such a way as to end regional disparities and to prevent the excessive growth of already crowded cities—But what the government is aware of and what it professes to be doing is often completely divorced from what is happening...
>
> The government is able to exercise direct control to some degree by use of its licensing procedures, but this can only prevent location not direct it elsewhere.

Government influence in the location of industries in India should, however, be viewed in two distinct phases: The first, the Nehru-era when the States' rights sentiment was not so intense and the Indian National Congress, as a united all-India political party, ruled almost all the States of the Union.

The second, the post-Nehru period when the sentiment for state autonomy, which was already rising in the last days of Nehru, became much more pronounced and regional political parties with somewhat exclusive territorial bases came to control several of the State governments. In the first phase, India could well be described as a unitary country with some federal features. But in the second phase, as the states' demands for autonomy was further reinforced, the country has progressively become what may be termed as a federal country with certain unitary features.

During the first phase, the effectiveness of the Centre in the sphere of economic activities was almost the same as during the British Indian regime. The regional governments were under the control of the same political party as was the national government, and the High Command of the Congress Party remained an effective regulatory organ. The flow of capital, labour, and commerce remained free, and inter-State boundaries presented no obstacle to the flow of men and materials. Economic investments continued to be located in areas most profitable to the investors. The central government despite its socialistic stance, did not greatly interfere with the growth of industries in the private sector.

Thus, as was natural under a somewhat *laissez-faire* economic policy, the regional ineqalities of the British period continued to prevail. As Table 6.1 shows, over eighty per cent of the licenses issued during the period 1953 to 1961 were granted to industrial establishments located in the areas of "third" and the "highest" levels of development. The lesser-developed areas experienced minimal growth of private sector during the period, so that "the government licensing controls can be said to have had very little value in promoting the development of these areas" (Roth, 1.970, p. 393).

During the earlier period of central dominance, the constituent states of India possessed no strong *raison d'etre* as separate economic units. But, with the weakening hold of the centre and the rising demand for States' rights, the nature of Indian federalism has changed considerably, and this has greatly affected the effectiveness of the central government as the agency of landscape change. Although, the flow of trade and capital is still as free as ever, the pattern of politics born out of the multi-ethnic character of the Indian federation has largely changed the ground realities over the past 40 years.

The British Indian provinces had possessed no legal autonomy and were largely artificial creations with which the people in general had little identification. Consequently, demand for state autonomy in India was originally quite weak. However, following the reorganisation of Indian states on the basis of languages, the state's rights sentiment greatly increased so that the state governments in India have increasingly been transformed into zealously guarded homelands, and as preservers of the rights and privileges of their co-linguists.

In a country with heavy pressure of population and prevailing unemployment, these rights permeate all sectors of economy. This has, in practice, placed considerable restrictions on the free flow of labour and capital. Entry of outside labour is discouraged by the states and local recruitment is insisted upon even in projects sponsored by the centre.

Table 6.1 : Licenses issued to Industries by State and Level of Development 1953 to 1961

State	Level of development				
	Total	Lowest	Second	Third	Highest
Andhra Pradesh	141	–	24	59	58
Assam	48	–	4	26	18
Bihar	146	11	30	13	92
Gujarat	377	–	3	84	290
Kerala	285	–	5	189	11
Madhya Pradesh	-86	3	9	8	66
Madras	499	–	–	–	–
Madras District	(140)	–	–	–	140
Remainder of Madras State	(359)	–	–	86	273
Maharashtra	1,357	–	–	–	
Greater Bombay	(1,071)	–	–	–	1,071
Remainder of Maharashtra State	(286)	2	26	20	238
Mysore	191	3	24	52	112
Orissa	65	27	26	12	–
Punjab	362	1	1	2	358
Rajasthan	62	3	8	5	46
Uttar Pradesh	368	33	16	80	239
West Bengal	927	–	–	–	4
Calcutta	(567)	–	–	–	567
Remainder of West Bengal	(360)	–	1	1	358
Delhi	134	–	–	–	134
Himachal Pradesh	3	1	2	–	–
All India	4,1	4	1	63	4,071

Source: Census of India (1961), vol. Part 1A (ii), 'Levels of Regional Development'.

Likewise, flow of capital is also adversely affected. Firstly, with the assumption of regional autonomy, the state governments have become rivals rather than partners, as they primarily were in the colonial period or soon after independence. Nowadays, the States, depending upon the political and economic philosophy of the party in power, try to attract industries by wooing investors by way of increased incentives. This could not have been possible if the country had a unitary rather than the federal set up. Secondly, following the decline of the Indian National Congress, Region-based political parties have gained control of government in a number of states. Some of these are averse to private enterprise and are in the habit of extracting political mileage out of industrial unrest. Naturally, enough this has resulted in the flight of capital from these areas, leading to adverse effect on progress toward a balanced economic development across the nation.

It is partly for this reason that the state of Kerala, which has off and on been voting Communists to power, has failed to attract any great industrial enterprises in the private sector. Calcutta in West Bengal, on the other hand, because of its advantages of historical momentum, and its economically strategic location had become one of the chief economic hubs of the country. It had continued to expand uninterruptedly after independence and the inauguration of the federation. In the post-Nehru era, following the installation of government by Communist-dominated coalition parties in 1960s, government-supported strikes and industrial disturbances became common. Industry had to put up with considerable losses in production and profits besides a good deal of harassment to the executive personnel. This forced entrepreneurs, to move away from West Bengal to safer though, less profitable locations for capital investment. Many large establishments shifted their headquarters to centres outside West Bangal and others were contemplating similar moves. As the Overseas Hindustan Times commented (13 June, 1970):

> It is an ill-wind that does nobody good. Owners of vacant flats in Delhi have been finding that out since the Naxalite scourge began uprooting business houses from Calcutta. Most of the business which decided lo

> pull out and cut their losses have shifted to the Capital. Here, they plunged into a desperate hunt for office accommodation and homes for their executives. By last week, a typical modern two-bedroom suite with lounge-dining room, kitchen and bath had jumped from Rs. 400-450 a month to Rs. 550-600 a month and ordinary house-hunters had been squeezed right out of the market... businesses had begun fleeing tempestuous Bengal, a while before peasant-landlord clashes in the countryside and naked thuggery in the city brought law and order to its knees. The advent of President's Rule failed to stem the flood.

Things improved in West Bengal following elimination of militant Naxalism in the 1970s leading to restoration of the old pattern. In the 1980s, a similar fate awaited punjab and the state of Jammu and Kashmir, with almost similar impact on the housing market in and around Delhi.

7

THE NATION-STATE IN POLITICAL GEOGRAPHY

INTRODUCTION

Although, the concept of State as a sovereign entity has been the fundamental subject matter of modern political geography right from the start, theoretical discussion on the nature of State remained a neglected topic in political geography. Ratzel did, of course, propound a meaningful theory of the State as an organism, but his concept of State was biological rather than political-scientific in nature and inspiration. Owing to this, Ratzel neglected considerations of ethics and morality. The State was conceptualized more like a mechanical contraption devised to fulfil the national urge for spatial expansion.

The rise of German geopolitics, the subsequent rejection of Ratzelian thought as the basis for political geography, and the emergence of an alternative concept of political geography as chorology, transformed the concept of the State from a purposeful political organization to a mere territory. It was distinguished from other areas only in the sense that it was defined in terms of political jurisdiction. The State began to be looked upon simply as an areal object, and, according to *Hartehorne* (1935), the primary task of the political geographer was to study its morphology.

Focus on the State as a purposeful organization received attention for the first time since Ratzel, in Hartshorne's (1950) "Functional Approach in Political Geography", wherein he made a forceful plea that political geographers should direct their attention to the State's functions pertaining to "the creation and maintenance of law and order", and to securing "supreme loyalty of the people in all its regions, in competition with any local or provincial loyalties, and in definite opposition to loyalty to any outside State-unit". The Functional approach promoted a theory of State which was very much in line with the post-war concern in political and social sciences with the functioning of the State as a political/economic organization. However, partly because of Hartshorne's insistence on chorology and exceptionalism, and partly because the concept was introduced at a time when the quantitative revolution was transforming human geography into a science of spatial analysis focused upon the study and analysis of abstract spaces, with a view to developing pattern laws. In such a situation politics and political theory had no role to play. Little attention was given to the theory of the State as a purposeful organization. However, the picture started to change in the 1970s. The call for social relevance in human geography in the early years of that decade, generated considerable interest in topics such as systems of government—their structure, functioning and socio-political consequences (Dikshit, 1971, 1975)—and the role of State as an instrument of distributive justice (Smith 1977). By the end of the decade interest in the theory of the State had considerably increased, leading to good deal of deliberation on the nature and purpose of the State as a form of social organization (Dear and Clarke 1978, dark and Dear 1984, Short 1982).

THE IDEA OF THE STATE : A HISTORICAL PERSPECTIVE

The State is so all-pervading in our day to day life that the point is very often missed that the idea of the State is a very recent development. Its antecedents do not go beyond the modern phase of history. Modern history covers the period that began with the renaissance and the reformation in the fifteenth and sixteenth centuries, and moved forward through the Scientific revolution of the seventeenth century and the Enlightenment of the eighteenth century, to culminate into the Industrial revolution of the 1780s, followed by the French revolution of the 1789. The ground had thus been prepared for the "European Century" (1815 to 1914), the century of liberty and progress during which European ideas, techniques, and political rule spread to the far corners of the world. The Industrial revolution gave unprecedented wealth and power to a few leading West European countries, and by creating a huge divide between them and other countries, compelled the latter to accept the ideas and technology of the West.

Raising labour productivity through latest techniques provided a new material basis for civilization, altering the organization of human societies and their relationship with nature. This transformation was closely linked with the rise of a capitalist economy and the "bourgeois revolution". By changing social and political relations in society, this revolution led to the rise of the modern State as an authority which was entirely secular, representing supreme public power within its national, territory. The public power of the State was not identical with the personal authority of any single individual. The public power of the State began to be organized through permanent agencies and institutions, creating favourable conditions for the rise of free markets in labour, goods, and land, and consequently, a "civil society"—conceived as a private realm where individuals were free to pursue private interests.

Two central ideas in the modern concept of State are:

(i) That it is a centralized power created to overawe all other powers within its territory. The State power is exercised through secular agencies like the bureaucracy, the judiciary and the army,

(ii) The power and organization of the State is founded on mutual consent—on an active inter-relationship between those who direct it and those who are subject to it. Thus, the State becomes a kind of civil association, and the government of the State an instrument of society. This leads to the notion that

> if the State is a "true moral community", then the laws that it establishes and enforces must be binding on all its members. Under certain conditions, all must have an obligation to obey the laws that are passed. At the same time it follows that if those conditions are not

> met, the individuals are released from their obligation to obey the commands of the agencies of the State...The sovereignty of the modern State is proclaimed but this means not just that the State is recognised as power, but as a power that must be made legitimate, so that it is transformed into authority (Gamble 1981, p. 49).

With the enshrining of the State as a secular and independent institution acknowledging no authority higher than itself in its own sphere of activity, the social basis of the State went on widening to include all citizens irrespective of class and caste distinctions.

As a civil association the modern State has not been simply a negative guarantor of civil society and its interests, but an active creator and moulder of that society, and has come increasingly to be used to plan societies along certain preconceived lines, wrenching societies away from traditionalism toward modernisation and development (Gamble 1981, p. 49).

There are two broad perspectives on the State. One is called, the democratic-pluralist view which regards the State as a neutral arbiter between diverse competing interests in the national community which are engaged in promoting their interests through formal and informal political channels (Dahl, 1956). This is known as the *democratic-pluralist* view of the State. However, as many commentators have noted, the actual dynamics of the State politics is far from a contest between equals. State policies are often decided by the more powerful groups within the society by influencing decisionmakers who may be sympathetic to their ideology and economic philosophy. Secondly, State politics in our times has reached a stage where the individual voter and small groups have ceased to matter. Public policy is increasingly dominated by federations of organized labour and dominant capitalist interests. As a result, the democratic pluralist State is fast moving toward a corporate State (Galbraith, 1967; Middlemass, 1979). This view is called *imperfect pluralism*.

The second broad perspective on State is the radical, Marxist, perspective which rejects the theory of the State as a neutral arbiter. According to this view, the State is not above the class struggle and social conflict in society, it is part of them. According to the Marxists, all societies are class societies in which one class dominates other classes, and the State is the means and support (*i.e.*, the instrumentality) for this domination. This view, termed as the *instrumentalist approach*, is as- sociated with the ideas of Miliband (1969). An alternative view, called

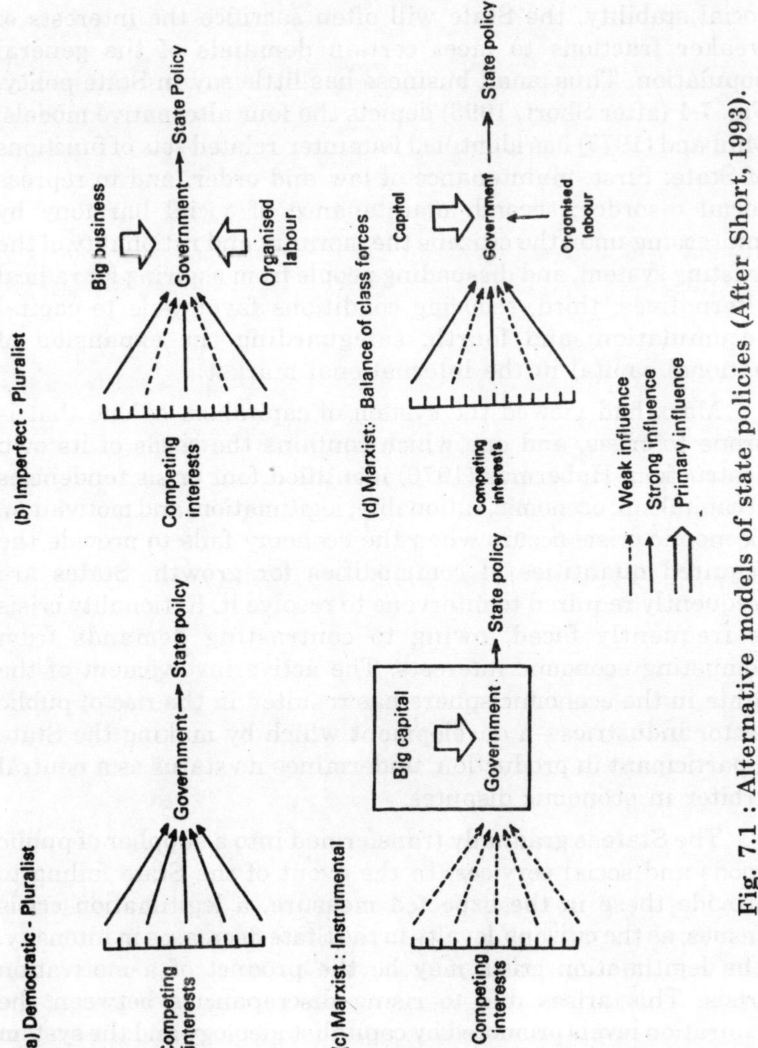

Fig. 7.1 : Alternative models of state policies (After Short, 1993).

the balance of *class forces approach*, takes a more complex view to the situation. The State is viewed as an arena of competing interests, and not simply the vehicle of class domination. The State represents the power of the capitalist class, but the various fractions of this class have conflicting interests. To achieve social stability, the State will often sacrifice the interests of weaker fractions to meet certain demands of the general population. Thus small business has little say in State policy. Fig. 7.1 (after Short, 1993) depicts the four alternative models. Miliband (1977) has identified four inter-related sets of functions of State: First, maintenance of law and order, and to repress social disorder; second, maintenance of social harmony by impressing upon the citizens the morality and rationality of the existing system, and dissuading people from aspiring for radical alternatives; third, securing conditions favourable to capital accumulation; and fourth, safeguarding the expansion of national capital in the international market.

Marx had viewed the system of capitalism as one that is prone to crisis, and one which contains the seeds of its own destruction. Habermas (1976) identified four crisis tendencies of capitalism: economic, rationality, legitimation, and motivation. Economic crisis occurs when the economy fails to provide the required quantities of commodities for growth. States are frequently required to intervene to resolve it. Rationality crisis is frequently faced, owing to contrasting demands from competing economic interests. The active involvement of the State in the economic sphere has resulted in the rise of public sector industries—a development which by making the State a participant in production, undermines its status as a neutral arbiter in economic disputes.

The State is gradually transformed into a supplier of public goods and social services. In the event of the State failing to provide these in the expected measure, a legitimation crisis ensues, as the citizens' loyalty to the State decreases in intensity. The legitimation crisis may be the product of a motivation crisis. This arises due to rising discrepancies between the aspiration levels promoted by capitalist ideology and the system of education, on the one hand, and the evolving socio-cultural system, on the other. The source of the crisis lies in the fact that the capitalist system of political organization has, over time, given rise to social groups and ideas which are not consistent with its own ideological basis.

The State in modern capitalist societies must perform two basic (but somewhat contradictory) functions: it must ensure conditions of capital accumulation, and it should continue to strengthen the basis of its own legitimation by retaining loyalties of the masses. (O' Conner, 1973). The two sets of functions involve the State in two different sets of financial expenditures. Capital accumulation calls for the generation of social capital, which involves capital expenditure in building and maintaining infrastructure for continued economic growth. The legitimation function requires social expenditure involving expenses on social welfare, internal order and external defence.

CRISIS OF THE TERRITORIAL STATE

That the traditional notion of territorial sovereignty of States has been greatly undermined with the development of transport technology, weapons of war, and a unified international economic system, has been on the research agenda of the social scientists for some time (*e.g.*, Hertz, 1957), and has received a good deal of attention by students of the State as a system. The advent of the nuclear age in 1945, conclusively proved that the States could no longer maintain territorial sanctity of their space, and perform their basic function of defence against outside agencies (Deustch, 1981). Domination of national economies by the context of the world economic system has greatly reduced the scope for autonomous action in the economic sphere (Holland, 1976). And the number of problems requiring international cooperative effort are progressively increasing—problems such as environmental crisis, and the widening rich-poor gap. Brown (1973) drew attention to the rapid erosion in national sovereignty resulting from the growing economic interdependence in the international system. This aspect of the emerging world scenario was the subject of focused attention in a number of studies in the early 1970s (Murray, 1971; Goldstein, 1974; Holland, 1976).

There was also a good deal of discussion on the role of multinational corporations in reducing' the effectiveness of

individual States in the management of their national economies. The multinationals are not affected by changes in interest rates since investment comes either from retained earnings or through international borrowing. These companies also have the advantage of intra-company trading across international borders, which helps them to avoid high taxation in the country of production (should that be the case). As Holland's (1976) study showed, the operation of multinational companies have undermined the State's authority in planning industrial growth towards a balanced development. One of the methods employed by States to achieve this objective has been through incentives for industries moving into areas of depressed economic growth and creating disincentives for further growth in the already developed parts. These measures do not work with the multinationals. If faced with disincentives, they can easily transfer their expansion plans to some other more favourable location outside the concerned State.

There is no denying the fact that States today "are not sovereign, in the meaningful sense of the term, and are not self-contained, are not impermeable" (Scott, 1967, p. 18). But despite this increasing trend toward nullifying the sovereignty of the territorial State, the international system continues to be sub-system dominated. National governments remain the most important mediating agencies between individuals at home and others abroad (Singer, 1969, p. 27).

THE CONCEPT OF NATION

The term nation denotes a community of people bound together by a general sense of belonging caused by common sharing of historical memories, tradition, language, religion, culture, and political State, Nation and the Nation-state 195 goals. Although it is territorial in character, it does not involve territorial sovereignty even though the concept of nation is supremely a political concept. One of the best definitions of this term was given by John Stuart Mill (1861) who defined it as "A portion of mankind united among themselves by common sympathies which do not exist between them and any others—which makes them cooperate with each other more willingly than with other people, desire to be under the same government, and desire that it should be government by themselves exclusively". In other words, a nation denotes a group of people with a strong sense of belonging—a group whose members place loyalty to the group as a whole, over any conflicting loyalties.

It must be noted, however, that although the term nation implies common culture, common symbols, and a particular view of the world which is distinct from other world views, "what makes a nation different from other cultural groups...is that one of the symbols associated with its values and attitudes is a particular piece of territory" (Lanyi and McWilliams, 1973, p. 40). The concept of nation, therefore, is essentially a geographical (*i.e.* territorial) concept.

The state of nationhood is an advanced stage in the ongoing process of political and social integration. This stage is reached through a gradual process of the broadening of social communication between societies as they evolve into a community, then into a people, and finally into a nation. In modern literature on political integration (of which nation-building may be treated as a sub-type), a society is defined as "a group of persons who have learned to work together"; a *community*, as "a group of persons who are able to communicate information to each other effectively over a wide range of topics", and a *people*, as "a group of persons with complementary communication habits". And in this context, a *nation* may be defined as "a people which has gained control over some institutions of social coercion, leading eventually to full-fledged nation-State"; and *nationalism* is the "preference for the competitive interest of this nation and its members over those of all outsiders in a world of social mobility and economic competition, dominated by values of wealth, power and prestige, so that the goals of personal security and group identification appear bound up with the group's attainment of these values" (Deutsch, 1953). But "Nationalism, like any other form of loyalty is a matter of degree—a given people at a given time may be more or less a nation, while none fully approximates to the ideal type" (Rustow, 1968).

Defining "the forces of cohesion for the integral community" (*i.e.* a nation), Reinhold Niebuhr (1959, p. 149) enumerated the following:

> common language and a sense of ethnic kinship, geographic unity and continuity, a common historical experience and frame of political thought, a common area of economic mutuality, and sometimes, the fear of a common enemy.... Any of these forces may be defective, but they cannot all be defective if the unity of the nation is lo be preserved. A common religion was usually regarded as an equally important prerequisite until modern religiously pluralistic nations...refuted the theory.

Regardless of its basis, cultural cohesiveness is transformed into nationalism only when the ethnic-cultural identity becomes organized for the achievement of political demands. The first coherent manifestation of modern nationalism occurred in seventeenth century England, when the authoritarian tradition on which the Church and the State rested, was challenged by the English Revolution in the name of liberty of man. It was the liberal nationalism of England's Glorious Revolution that inspired the French philosophers. The French Revolution was guided by the English model of constitutional liberty and limited government (though as the Revolution progressed, the absolute sovereignty of the king was replaced by the absolute sovereignty of the people). In a sudden burst of national solidarity, in August 1789, all geographic and class barriers had fallen apart.

The Declaration of the Rights of Man and Citizen established the basis for a nation of free individuals protected by law. It was through the French Revolution that the English ideas about liberty and nationalism were carried to other parts of Europe. They were absorbed and transformed into the general consciousness of the eighteenth century western mankind, and subsequently gained universal acceptance as a "model for awakening liberal thought" (Kohn, 1955, p. 19). During the nineteenth century the concept spread all over Europe; the second French Revolution of February 1848 was followed by revolutions in most European capitals, and in the twentieth century, it became a worldwide movement: "the first reality of the twentieth century" (Senghor, 1959, p. 45).

The marxist scholar Gellner (1964) interpreted the rise of modern nationalism as a product of the geographical inequality of economic growth and development. The effects of the Industrial Revolution in England in the 1780s had spread progressively outward leading to transformation in agriculture, and to steady growth of industrialization, first in Germany, then in Italy, then in the Balkans and Eastern Europe. The geographic unevenness of the process of development among nations was shaped by the nature of resource endowment and the differential speed of the diffusion of capitalist relations of production. The spatial differentiation produced by these processes provided the context for the later rounds of capital investment. Thus, nationalism was not only a reaction to perceived regional inequality but also an attempt to guide economic development.

It was the ideological rationale for the drive from semi-periphery to the core. Drawing on the line of argument initiated by Gellner, Nairn (1977) has further advanced the theory of nationalism as a reaction to spatial unevenness in economic development. He argues that the elite mobilized the masses against the new style of development which was undermining traditional values. This was accomplished by invoking cultural patriotism through the revival of old historical memories and traditions. The national patriotism so generated was motivated by a competitive spirit to drive the country forward to a core status in the international economic system, and to build a better economic future. Nationalist movements are, in this sense, both backward and forward looking at the same time.

STATE-FORMATION AND NATION-BUILDING

Nation is viewed here in the context of the nation-State. At the more primitive level, the concept of nation represents a feeling of kinship. As the people in a State mature in political experience, their feeling of belonging together becomes less dependent on ties of language and "race", and more on a common adherence to particular political concepts, ideals and institutions. The distinction between majority and minority communities dwindle into insignificance, as the people learn to devote their ultimate loyalty to the State and its institutions.

Thus, *nation-building may be viewed as the processes and the devices through which the sentiment of loyalty for the nation-*

State is fostered, and fissiparous and divisive tendencies are pacified and eliminated (Dikshit, 1976). In other words, nation-bulling may be viewed as the process of achieving political stability within the sovereign nation-States through the mechanics of active and free participation by its nationals in the political process of the State. In this context, nationalism may be defined as the "justifying ideology of the nation-State"; a system of "symbols of justification for the acts of a State...symbols by which... the State-organized cohesiveness of a nation is advanced and justified" (Gerth & Mills, 1953, pp. 198-199). In the terminology of political geography, nation-building consists of the devices through which the "State-idea" is strengthened.

Whereas nationalism is the political expression of nationhood, the nation-State is the politicogeographical expression of nationalism. Ever since the French Revolution "nationalism has been the main spiritual and emotional force cementing all elements of Statehood in the nation-State... Wherever the nation-State was a reality, nationalism buttressed and reinforced the State; where it remained an aspiration, nationalism endangered the existing multinational units" (Frankel, 1964, p. 14). As a host of political geographers, since the time of Ratzel, have noted, a historical perspective on nation-States reveals that a modern State is likely to show the greatest stability and permanence when it corresponds closely with the nation. In such a case, the State becomes the political expression of the nation, the mechanism through which the welfare of the members of the nation is safeguarded and its identity protected.

Nation-building (or political integration in the context of the nation-State) refers essentially to the process of bringing together, culturally and socially discrete groups into an organic (*i.e.*, fully functioning) system, and the establishment of a strong national identity inseparably bound with the territory of this system. There are four important aspects of this process: First, progressive development among the constituent political communities of a deep and unambiguous sense of identity with the State and its ideals. Secondly, the broadening and intensification of social communication among the constituent communities and regions. Thirdly, elimination of social, economic, and psychological commitments and preparing the people for new patterns of socialization and behaviour. Lastly, improvement of transport facilities by minimizing physical distances and thereby facilitating a greater flow of goods and services between the different regions of the State. This, alongwith the development of mass media and the growth of urban centres, pull the different components of the State's population together, and thereby, give rise to common norms of behaviour that reinforce a sense of unity and national identity.

Thus, nation-building and national integration refer "specifically to the problem of creating a sense of territorial nationality which overshadows—or eliminates—subordinate parochial loyalties" (Weiner, 1965). Nation-building is, therefore, essentially a politico-geographical problem requiring a geographical perspective in its solution. It involves merger of local place-loyalties into a larger territorial loyalty, until the entire territory of the State is transformed into a single, fully organized, place-entity—the fatherland of all citizens, claiming their ultimate loyalty. *Nation-building is, therefore, essentially a study in the sociology and politics of place.*

GEOGRAPHICAL PERSPECTIVE ON STATE-FORMATION AND NATION-BUILDING

Geography is sometimes described as "a discipline in distance". Distance is defined here in its widest connotation, and includes socio-economic and all other distances. In his attempt to measure the earth in terms of geographical distances, the geographer is inevitably led on to the consideration of economic and social distances. In doing so he becomes more fully aware of the total number of factors at play in creating the spatial patterns he may be investigating. Thereby the geographer is able to give a more comprehensive explanation of the phenomenon in question.

> In doing so he realises... how, and how much, geography has become the measure of man, measuring the reach of his success or the degree of his failure as man has struggled, to extend his ideas, his ways, his numbers and his control over the world around him (Watson, 1955).

And as Watson added, in his attempt to study the earth's surface (or any portion of it)

> the geographer must go on to those ultimate distances, traced in history and expressed in society, where the people through their customs, idiosyncracies, and aspirations, indeed in all the manifestations of their being, change and master the earth, to make of it but the imprint of their own purposes and soul. Geographical distances are in the last resort, the distances of human spirit itself.

This concept of geography is useful in underlining the process through which the multifarious diversities within the State get transformed into unity through gradual reduction of distances—physical as well as human—between the constituent localities and regions.

Geography is sometimes also described as the study of regions and areal aggregates. The study of regions of whatever sort (physical, political, cultural) requires focus on the areal aggregate, rather than on certain elements of the totality in isolation. In the study of nation-building, we are inevitably concerned with a special type of territorial unit, one defined in terms of political control exercized through free participation of the citizen in a democratic fashion.

Ratzel defined the State as a section of land and a group of humanity organized as a single unit in terms of a particular distinctive idea—the State-idea. Those States are strongest, Ratzel had concluded, "in which the political idea of the State fills the entire body of the State, extends to all its parts", for in such cases, the entire State becomes a single nation. In the well-developed modern State, politically active people in all regions are conscious of their loyalty to the State and have some common understanding, of what that State means to them. In such cases, in terms of their geographical spread, the nation and the State will be identical.

It is with the concept of State-idea that the geographer ideally starts his analysis of the State. His first concern is to determine the area to which the idea applies, and then the degree to which it operates in the different regions of the territory occupied by the State. On this basis, a distinction is made between those regions that form integral parts of the State-area in terms of its basic idea, and those parts that may be held under control in the face of indifference or opposition on the part of the regional population. In this way the geographer determines the areal correspondence of the State and the nation, and identifies the regions in which the national character is partial rather than complete. In this way he can determine the degree to which the diverse regions of the State constitute a unity. He may thereby isolate areas of weakness and strength from the view-piont of national unity.

Since few State-areas are homogeneous in respect of all the essential factors promoting national unity, the fundamental problem of nation-building in every case is to construct a *raison d'etre* that will enlist the loyalty of the diverse regional groups that may be having different associations and ideals. Each State "must formulate a set of purposes to which the citizens of the State can subscribe,... and which is sufficiently distinct from the purposes formulated by other States. Such a statement of purposes, with which the citizens of the State can identify themselves, constitutes' the State-idea" (James, 1968, p. 33). As Preston James noted:

> The State-idea is a complex of traditions, experiences, and objectives. It is made up of written history, folklore, stories of national heroes, religious beliefs, and the language and art forms in which these things are communicated. It is the body of literature, the painting, the architecture, the music which are distinctly national. And it is the characteristic economic, social and political institutions. The State is created to defend and develop the State-idea.

The State-idea is not always easy to identify, nor is it easily quantified. However, the attitudes and objectives forming the basic idea of the State can be measured and cartographically displayed.

Nationalism and nation-building are primarily concerned with the people constituting the State as a political community. It is necessary therefore that we identify the basic types of national societies on the basis of their geography, *i.e.*, on the basis of variations in the spatial patterning and areal associations of their chief identifying elements—ethnic, historical and others. On this basis two broad types of national societies may be recognized: Firstly, *Functional*, in which the different identifying elements are mixed together like strands of different colours in a marble cake. Secondly, *Federal* societies, *i.e.*, the ones in which some of the more important identifying elements possess marked regional groupings. The former type, possessing no durable bases for the demand for regional centres of decision-making, should, under normal circumstances, lend itself to a unitary form of governmental system. The latter, with its well-defined regions, epitomizing concentration of strong regional loyalties, should, under normal conditions, lead to a federal arrangement. The problem of nation-building in the two types of States are, therefore, qualitatively different.

The process of political integration primarily involves the emergence of a core area—an area where administration and economic capacities exceed the load of political demands (Whittlesey, 1939). In a durable nation-State there must be a compensating flow of advantages and sacrifices linking the various areal parts, and rival plans of integration should be eliminated. This is most effectively accomplished by reinforcing the socio-psychological impression in the minds of the peoples in the constituent regions that, in the ultimate analysis, their sacrifices are smaller than their gains (Dikshit, 1975, p. 198). By the very nature of their geographies, opportunities for rival regional plans of integration are, comparatively speaking, minimal in the case of functional societies, but considerably greater in the federal ones. In the latter, each constituent region signifying strongly identified assemblages of pieces of territory and groups of people, may possess latent potentialities of becoming a Nation.

This distinction between potentially functional and potentially federal structures is a point of fundamental importance that any student of State-formation and nation-building must keep in mind, and use as the starting point of his analysis. Detailed analysis of the processes and problems of nation-building in particular States, in the context of "the compensating flow of advantages and sacrifices" referred to above, should greatly benefit from the various forms of transaction-flow analysis. As Deutsch (1964, p. 51) wrote, "the study of quantitative densities of transactions is the first step towards estimating the degree to which people are connected with one another". (However, nation-building is essentially qualitative rather than quantitative in nature, so that its study calls for a focused place-specific humanistic approach.)

Such a spatial perspective on the problem of political integration, and the fundamental distinction between potentially federal and functional societies has so far remained completely neglected. This is unfortunate since, viewed closely, this distinction may prove to be as basic to the whole question of state-formation and nation-building, as the distinction between military societies (in which integration is achieved through compulsion), and industrial societies (which achieve integration through voluntary co-operation); or that between lower societies (which achieve integration through "mechanical solidarity" achieved through the penetration of such collective values as heredity and station in individual conscience), and higher societies (which achieve integration through the process of "organic solidarity" stemming from the differentiation of individuals according to the social functions they perform).

Index

H

U

V

W

❑❑❑